Praise for

THE TROUBLE WITH GRAVITY

"Gravity is fundamentally a mystery, writer Panek reveals in this beautiful and philosophical investigation of nature's weakest force . . . Readers will not emerge from this book with the answer to the question 'What is gravity?' — a so-far-unanswerable quandary — but they will gain many and varied insights from the asking." — *Scientific American*

"Panek takes evident pleasure in the whirl of new ideas. He has made a career out of explaining things that scientists themselves may barely understand." — *Washington Post*

"By confessing his own amazement, [Panek] invites the reader to share in the adventure. He does so with humility and humor." — *Undark*

"Highly recommended. Both accessible and enjoyable." — *Library Journal*

"Panek's inquisitive, fine-tuned narrative is full of character and, unlike many other books on physics, imbued with the friendly casualness of a coffee-shop chat. As such, it will delight both lay readers and serious students." — *Publishers Weekly*

"[A] fine popular primer . . . [An] expert description of the spectacular things that gravity does." — *Kirkus Reviews*

"With a sustained sense of wonder, Panek finds the roots of science in our myths and poetry, uncovering the provocative side of something we only think we know. His rigorous-but-readable book won't reveal what gravity is, but will challenge your view of the universe and our place in it." — **Apple Books, "Best Book of the Month"**

"In *The Trouble with Gravity,* Richard Panek acts as a guide, both amiable and erudite, through one of the most puzzling mysteries of the natural world. In explaining the various 'explanations' of gravity from classical to postmodern times, Panek draws us into a thoughtful meditation on the mythic, cultural, philosophical and, yes, scientific implications of what happens when a wet potato or a crystal vase slips from your hand." — **Billy Collins**

"I've long been a big fan of Richard Panek's writing. He is eloquent, smart, and a fascinating thinker, someone who is able to get me excited about topics that would have never even occurred to me. I respect and trust him — and am always eager to see what he will write next."
— **Maria Konnikova, author of** *The Biggest Bluff*

"Richard Panek moves with startling grace and economy through the intersecting realms of philosophy and physics, always asking the unexpected question. He has forced me to rethink my fundamental assumptions about gravity — and shown me how much we can gain by doing so."
— **Andrea Barrett, author of** *The Air We Breathe* **and** *Archangel*

"Without gravity, there would be no Earth, no humans, and no nonfiction books. Which would be a shame, because we'd miss out on Richard Panek's wonderful, entertaining work. Richard takes us on a vivid journey from the arctic to the tropics, from the human skeleton to the edges of the universe, filling our imagination with counterintuitive modern science and ancient philosophy. And all of this is delivered in buoyant, almost poetic, writing. So thank you, gravity and Richard."
— **A. J. Jacobs, author of** *The Year of Living Biblically*

"Gravity is a mystery — one of the greatest. It has baffled and teased humans since the dawn of history and perplexes us still. Richard Panek takes us on a journey that is original, brave, and ultimately very beautiful: a reminder that sometimes science isn't a solution but a search."
— **James Gleick, author of** *Time Travel: A History*

"A thoroughly researched tour of humanity's investigations of gravity through the ages, including the very exciting — but still unfinished — ones happening today."
— **Lisa Randall, professor and author of** *Dark Matter and the Dinosaurs*

THE
TROUBLE
WITH
GRAVITY

THE TROUBLE WITH GRAVITY

SOLVING THE MYSTERY BENEATH OUR FEET

RICHARD PANEK

Mariner Books
Houghton Mifflin Harcourt
Boston   New York

First Mariner Books edition 2020

hmhbooks.com

Library of Congress Cataloging-in-Publication Data
Names: Panek, Richard, author.
Title: The trouble with gravity : solving the mystery beneath our feet /
Richard Panek.
Description: Boston ; New York : Houghton Mifflin Harcourt, [2019] | Includes
bibliographical references and index.
Identifiers: LCCN 2018057178 (print) | LCCN 2019009530 (ebook) |
ISBN 9780544568297 (ebook) | ISBN 9780544526747 (hardcover) |
ISBN 9780358299578 (paperback)
Subjects: LCSH: Gravity.
Classification: LCC QB331 (ebook) | LCC QB331 .P35 2019 (print) | DDC 531/.14—dc23
LC record available at https://lccn.loc.gov/2018057178

Book design by Greta D. Sibley

Printed in the United States of America
DOC 10 9 8 7 6 5 4 3 2 1

As ever, for Meg, with love

You know, I think we should put some mountains here
Otherwise, what are the characters going to fall off of?
And what about stairs?

—Laurie Anderson

CONTENTS

GRAVITY:
AN INTRODUCTION

I fell.

I had been sitting in a chair for a quarter of an hour, killing time in a bookstore. I had selected from a nearby shelf a book that I thought might relate to the subject I was researching at the moment — I no longer recall what. I'd pushed my chair away from a communal table, crossed my legs, and opened the book to a random page. The section I turned to happened to be on gravity.

What I had thought about gravity before I sat down was what most people think about gravity, to the extent that most people think about gravity: It's a force of nature. What I learned in the next few minutes is that it's not necessarily a force. Isaac Newton thought of gravity less as a force than as something mysterious that *acts across* space. Albert Einstein thought of gravity less as a force than as something mysterious that *belongs to* space. And quantum physicists agree with both Newton and Einstein: Gravity is *something*.

I looked up from the book.

Is gravity, I wondered, just a word? A semantic convenience? A placeholder until a better noun comes along? Something that, until we know more about it, we've agreed to call the single cause of a universe full of effects?

At the time I didn't understand the nuances of these questions. Yet even I — someone who had experienced little curiosity about science until well into adulthood; who had cultivated a professional interest in science but had only a minimal educational background in it; who in fact had taken AP math courses in high school specifically to escape having to take science courses in college (a strategy that pretty much worked) — could tell that I had no idea what gravity is.

I gently closed the book and stood up — but not quite. My left foot, the one that had been resting flat on the floor for a quarter of an hour, was asleep. I am nearly six-six; I don't drop gently. My fall silenced the customers in my vicinity. Yet even as I righted myself, trying to hold on to the edge of the tabletop as well as my dignity, I realized that I had received a valuable reminder: You take gravity for granted at your own peril. I also recognized that I had discovered a new mission in life — the pursuit of an answer to a question I'd never thought to ask:

What is gravity?

❖

"What is gravity?"

I was on the phone with one of the visionaries behind an experiment that had recently validated a prediction Albert Einstein had ventured almost exactly a century earlier: Gravity makes waves. Like a pebble in a pond, a gravitational interaction of any kind sends ripples, only instead of disturbing water it distorts space. Not just through the waves emitted by a collision of two black holes, which is what this experiment had detected. The fact that gravitational waves exist on that scale means that gravitational waves exist on every scale, including the human. Raise your arm: gravitational waves. Shake your head: gravitational waves. Fall over in a bookstore: gravitational waves, only bigger, I'd like to believe. The announcement of the gravitational-wave detection, on February 15, 2016, was the kind of once-in-a-generation scientific event that you can accurately describe as having captured worldwide attention: headlines saturating newspapers, TV news, and the Internet; discussions dominating their subsidiary forums — letters to the editor, talking heads, comment threads, and email, including a message from an astrophysicist friend of mine that read, in its entirety, "Hellzapoppin'." Of whom better than Kip S. Thorne, a theoretical physicist who had been studying gravitation since the 1960s, who had co-envisioned the experiment in the 1970s, who had been helping to guide the project since the 1980s — and one of the key members of the team soon to run the table in world-class physics prizes: Special Breakthrough, Gruber, Shaw, Kavli, and, in 2017, Nobel — to ask what gravity is?

"That's a meaningless question," Thorne said.

Good enough. You can learn a lot from meaningless questions.

Here's one that cosmologists often get: *What came before the Big Bang?* One answer: That's like asking what's north of the North Pole. From that precise point, there is no north on the surface of the Earth. There's just the surface of the Earth, extending only south.

Meaningless questions help scientists understand what's wrong with the premises behind the questions — the unthinking assumptions that render the questions meaningless. In the case of *What came before the Big Bang?*, the common unthinking assumption is that space and time exist independently of the universe, an assumption that even physicists held until quite recently in the history of our species. Instead, cosmologists say, space came into existence in the Big Bang, all balled up, and the unballing-up we measure by what we call time.

What is gravity?, however, is not a question that scientists often get, primarily because non-scientists have no reason to even conceive of it. *What is gravity?* was probably not a question that Kip Thorne should have gotten from me, not because I hadn't conceived of it, but because I already knew the answer.

Twelve or fourteen years had passed since I'd fallen over in a bookstore. The pursuit I had begun that day — to discover what gravity is — had ended almost immediately, perhaps as soon as the next day, maybe even that same evening. It required only cursory research, a minute's worth of clicks on a keyboard. The answer: Nobody knows.

For a while I had initiated many conversations on the subject with a kind of convert's zeal. These conversations tended to fall into one of two categories.

Category One:

> ME: Nobody knows what gravity is.
>
> CIVILIAN: (Pause.) What do you mean, nobody knows what gravity is?
>
> ME: I mean that nobody knows what gravity actually *is*.
>
> CIVILIAN: (Pause.) Isn't it a force of nature?
>
> ME: Okay, fine — but what does that even mean?
>
> CIVILIAN: (Silence.)

Category Two:

> ME: Nobody knows what gravity is.
>
> SCIENTIST: That's right.

From a civilian perspective, *Nobody knows* can be confusing; in my experience, it often inspired a look that says *What's the catch?* No catch! But I understood the suspicion, and so my own answer to *What is gravity?* had changed over the years. It was no longer just:

Nobody knows.

It was:

Nobody knows what gravity is, and almost nobody knows that nobody knows what gravity is. The exception is scientists. They know that nobody knows what gravity is, because they know that they don't know what gravity is.

We know what gravity *does,* of course. In the heavens, gravity tethers the Moon to Earth, other moons to other planets, moons and planets to the Sun, the Sun to stars, stars to stars, galaxies to galaxies. On our own planet, we know that gravity is what planes have to overcome, and what idiots in bookstores need to beware. We all know what gravity does, and we know it without ever having to think about it.

Once I did start thinking about gravity, though, I couldn't not think about it. Once I started noticing it, I was noticing it everywhere. Which is only appropriate, because gravity *is* everywhere. It's not just present; it's omnipresent.

I was thinking about it when I stepped out of the shower or off the bus. I was thinking about it when I dropped a glass on the kitchen floor or as I trudged up a familiar neighborhood hill or if, stoner-like, I contemplated a chair.

A chair *is* because of gravity. It wouldn't exist if our bodies didn't need something to cushion our journeys to the center of the Earth. The seat, the bench, the floor, the bed, the step, the stoop, the stool, the terrace, and the tire: all because of gravity. The escalator. The elevator. So, too, the objects we use to keep not ourselves but other objects off the ground: the table, the desk, the nail in the wall, the nightstand and pedestal and counter and kitchen sink. Count the legs in my living room alone: sixty-seven (not including my own). Far out.

The grave: Of course! The ultimate cushion. Google *grave,* and there it is: from the Latin *gravitas,* suggesting seriousness, heaviness, weight.

But that's only half of the story. The other half didn't occur to me until long after my conversation with Kip Thorne, yet it came to me because I couldn't stop thinking about what he, after a long pause on my end, had said next: "What do you mean by what gravity 'is'?"

He was right: "Is" is a big word. Big enough to obscure the unthinking assumption that rendered my question meaningless: that *what gravity causes* was the whole story. It wasn't. The other half was *what causes gravity*.

I'll give away the ending to this book: I still won't know what gravity is. But along the way I will try to discover what my meaningless question means. I'll look at the myths we tell ourselves to help us make sense of our place in the universe: anchored to Earth, and wishing we weren't. I'll round up the usual suspects — Aristotle, Newton, Einstein — to try to figure out what they were trying to figure out about gravity. I'll think about why, in the past couple of centuries, we've come to accept seeming absurdities — distortions in space and time, black holes, the Big Bang — even though we haven't begun to understand the concept of gravity upon which their existence depends. I'll consider what gravity causes, what causes gravity, and why those questions *aren't* meaningless — and might, in fact, give life meaning.

But then, the question of what gravity is — or "is" — has never been strictly about physics. It's always also been metaphysical; it's always also been philosophical. How we think about gravity — *whether* we think about gravity — has directed civilization for centuries, millennia even, back before we had a word for what we one day would try, and so far fail, to define. Gravity isn't just something that guides our every negotiation with the material world. It's something even more mysterious. It's something that's evident in the creation of creation myths, the elisions of religion, the effects in an IMAX spectacular. It's something we reflexively incorporate into the subtlest recesses of our civilizations and our psyches. It's something that has defined our scientific conception of the universe, even the multiverse, and it's something that has defined our conception of ourselves. It's the greatest ghost in the grandest machine.

"Oh, the *philosophical*," Kip Thorne said. "In *that* case —"

GRAVITY IN OUR MYTHS

In the beginning were the heavens and the earth. (You could look it up.) Then came light and dark and, with them, day and night. Soon followed the beasts of the earth and the fowl of the air. What wasn't in the beginning, at least not explicitly, was whatever was creating this division. Still, that "whatever" was implicit in those binary distinctions. That "whatever" defined, with the exactness of a razor, the most fundamental divide of all: the horizon.

But you don't have to take the Judeo-Christian tradition's Word for it. Just ask the Celts: "In the beginning, Earth and Heaven were great world-giants." Or the Wulamba, an Aboriginal people in northern Australia: "In the beginning, there were land and sky." Or the Ngombe of Zaire: "In the beginning, there were no men on earth. The people lived in the sky."

Which is not to suggest that what was in the beginning was always earth and sky. "In the beginning," says one Chinese myth, "there was Chaos." In the beginning, according to another Chinese myth, "was the great cosmic egg." "In the beginning," say the Bush-ongo, another people of Zaire, "in the dark, there was nothing but water." "In the beginning," a saga from the Indian subcontinent says, "this world was merely non-being." Another saga from the Indian subcontinent, however, begs to differ: "How from Non-Being could Being be produced? On the contrary, in the beginning this world was just Being, One only, without a second."

Yet even creation myths that don't explicitly begin with earth and sky immediately find common ground (and air) with the creation myths that do. "In the beginning was Chaos" — but then: "Out of it came pure light and built the sky. The heavy dimness, however, moved and formed the earth from itself." Same with the great cosmic egg: "P'an Ku burst out of the egg, four times larger than any man today, with an adze in his hand with which he fashioned the world . . . He chiseled the land and sky apart." Even mere Non-Being, however logically problematic, quickly got where it had to go: "It developed. It turned into an egg. It lay for the period of a year. It was split asunder. One of the two eggshell parts became silver, one gold. That which was of silver is this earth. That which was of gold is the sky."

The stories continue from there, following narrative paths as varied as the symbols attending the birth of the universe — eggs, existence, water, whatever. Yet they share a trajectory, not least because they emerge from a common origin, at least to judge from the unanimity across cultures and continents, seas and millennia —

what one of the most influential mythology scholars of the twentieth century called "the primeval pair": earth and sky.

In the beginning? *Down here. Up there.*

Gravity is as old as the universe: So says physics. The *story* of gravity, however, is older than the story of the universe. When the earliest storytellers — whoever they were, whenever they were — decided
to tell the story of the universe, the concept of gravity didn't exist,
let alone the word for it. The investigation into why things fall didn't
begin until a couple of millennia ago, and the idea that an actual
tangible *something* might be the cause of all that falling would have
to wait until the seventeenth century. But back when storytellers
were first starting to think about how we got here, and they needed
to clear the landscape of everything except what they believed was
essential to a universe, they were already operating under the influence of an unthinking assumption that was working its magic on
their imagination. Earth and sky; *down here* and *up there*: The need
to make those distinctions — and the desire to erase them — would
make no sense without gravity.

All stories have to start somewhere, and almost all stories start
in the middle. They deposit us in a setting — a when and a where —
that leaves us wondering how the story got to this point. How did
these characters come to find themselves occupying this particular
place at this moment in time?

Creation stories, however, are an exception. They don't start in
the middle. They begin in the beginning. The *when* part of their

setting is a present without a past. "In the beginning" isn't just a variation on "Once upon a time" — an arbitrary *now* with which to start a story. It's the ultimate *now,* the one before which no other *now* exists. Chaos or the contents of the cosmic egg or some other amorphous state of potential being might be present, but the *story* of that universe is not.

And then, it is.

Creation happens — not necessarily the creation of the universe, which might already exist in the form of Chaos or the contents of the cosmic egg or some other amorphous state of potential being, but the creation of the universe that emerges from Chaos or the cosmic egg: the universe *as we know it.* The one we inhabit. The *where* of which we inquire: What is this place, and what is our place within it? How do we stand in relation to the whole?

For a start, we do just that — stand. We do not, alternatively, swim beneath the surface of the ocean, ignorant of air. Nor do we float in the wine-dark vastness between Earth and Sun or between star and star, regarding each with equal indifference or equal curiosity. We do not oscillate in the strobe-lit vastness between electron and nucleus or between atom and atom. What we occupy instead is one-half of the primeval pair, a surface we can't help identifying with, whether or not we know we do.

Not: We don't knowingly identify with the surface of the Earth because we don't think of ourselves as occupying it. We identify with our surroundings, because we inhabit them — earth, air, fire, water. But we don't identify with whatever it is that anchors us to our place in the universe, because *it* inhabits *us.* "Then the Lord God formed man from the dust of the ground": Just like that, Genesis put us in our place.

Paradise, our place could have been. Paradise it was, for a while, in many of our stories. Maybe we emerged in the sky, in a garden, or somewhere in between. We were the first of Africa's Luba people, sharing the celestial realm with deities, basking in our common immortality. We were Adam, naming the beasts at God's behest, not needing food or clothing or shelter. We were the first person among the Altaic of Central Asia, pairing off with God, the two of us skimming the surface of the primordial ocean in the guise of black geese. All was right with the world.

And then it wasn't. Something went wrong. We know that something went wrong because, well, just look around: not only food, clothing, shelter — those, we could live with — but disease, predation, death.

Having arrived at the answer to when and where our creation myths take place — "in the beginning"; "in a place with a *down here* and an *up there*" — we found ourselves confronting a new question: *Why are we down here and not up there?* Why are we at a great remove from whatever our original state might have been, whether in the garden next door, in the salty air above the sea, or in any other kingdom of perfection? Read enough of the world's myths and the collective plaint is actually rather touching, one species' *cri de coeur*: "What did we do to deserve this?"

Plenty, is the consensus.

In some myths what we did was banish our creator. Squeezed between our parents, suffocating within their selfish embrace, we pried Sky Father from Earth Mother by hunching our shoulders so powerfully that we forced him upward. Or we ripped apart the sinews and tendons their bodies shared. Or we squirmed until our father scrammed, dragging his wounded pride behind him. Then again, if he was merely nearby rather than smothering us, we irked him by tearing shreds from him to make our clothing, or wiping our hands on him, or throwing dirty wash water at him, or poking him in the eye with a pestle. If he collected us along with our siblings and exiled all of us back into our mother's womb, we waited for just the right moment, then emasculated him with a sickle.

Just as often our creator banished us. Basking in our celestial realm, we grew bored by immortality and bickered loudly with one another, like bad neighbors, practically daring our landlord to evict us. We ate from the Tree of Knowledge of good and evil. We uprooted the Great Turnip.

Either way — whether we banished our creator or our creator banished us — we were down here, God or god or gods were up there, and that's just the way it would have to be.

Would *have* to be?

Place mountain upon mountain upon mountain. This was the strategy of the brothers Otus and Ephialtes, who wanted to make war with the gods on Olympus.

Stack log upon log upon log. This was the strategy of Kamonu, seeking Nyambi's home in the sky.

Lay brick upon brick upon brick. This was the strategy of the descendants of the survivors of the Flood who had settled in the city of Babel.

But none of these strategies worked, and we know that none of them worked because, well, just look around. We're here, asking yet another question, not *Where are we?* or *Why are we* here? but *Why are we* still *here?*

We're still here — *down here* — because we couldn't get *up there*.

Because even though Otus and Ephialtes were nine fathoms high and nine cubits* around at the chest, they were only nine years of age and no match for Apollo, who spied their approach and killed them. Because the weight of Kamonu's pillar of logs was too great and the tower collapsed. Because God said, "Let us go down and there confuse their language, that they may not understand one another's speech." Staggering from mutual miscomprehension, they scattered themselves "over the face of all the earth."

As for flying, forget it. Take the myth of Icarus. The artisan Daedalus and his son escape prison by flying away on wings of feathers bound by wax, but Icarus disobeys his father's caution about flying too high, and the sun melts the wax, and Icarus plummets to the sea.

You might assume† that the moral and the meaning are pretty straightforward. The moral: Don't fly too close to the sun. Meaning: Don't be too full of yourself.

That's one interpretation. But it's an interpretation that fails to take into account the other half of Daedalus's caution: "Icarus, my son, I charge you to keep at a moderate height, for if you fly too low the damp will clog your wings, and if too high the heat will melt them." So the moral is . . . don't fly too close to the water, either? Meaning . . . all things in moderation?

---

* Fifty-four feet and 13.5 feet, respectively.

† I did, anyway.

But then Daedalus continues: "Keep near me and you will be safe." So . . . fly close to your father/obey your elders?

Palaephatus, the (probably pseudonymous) author of a manuscript that reimagined — or, more accurately, *de*imagined — the Greek myths of his day, doubted the veracity of the story itself, let alone the moral or the meaning: "It is impossible to think that a human being flew, even with wings attached. What happened was this" — and Palaephatus described Daedalus and Icarus boarding a small boat that skimmed the waves at such great speed that, to its pursuers, it "seemed to be flying." Then the boat capsized.

What someone else might read as an allegory, Palaephatus saw as an affront to logic.* But he got at least one thing right: In his telling of the story, the emphasis falls equally on father and son. Palaephatus recognized a balance that's missing from the many metaphorical interpretations of the tale, the ones that emphasize the tragedy of Icarus over the triumph of Daedalus. The myth — whatever its moral or meaning, or even its adherence to the facts of nature — is

---

* Palaephatus was full of these literalist interpretations. Centaurs? They couldn't have existed because "horse and human natures are not compatible, nor are their foods the same; what a horse eats could not pass through the mouth and throat of a man." Similarly, minotaurs: "It is impossible for one animal to make love to another if their genitals do not conform . . . Nor would a woman tolerate being mounted by a bull, nor would she be able to bear an embryo with horns." Maybe Palaephatus was trying to correct a common belief that myths were completely factual; more likely, he was a pedant pointing out the obvious.

indeed a tale of a boy who fell to Earth. But it's also the tale of a man who flew like a bird.

Until he didn't. Until, like a bird, he coasted back to Earth. The implicit inspiration behind this myth wasn't *Where are we?* or *Why are we down here?* or *Why are we still down here?* It wasn't even a question. It was a declaration of fact: We mortals are down here — on the face of all the earth — and down here we mortals have to stay.

The soldier awakens on a pyre — though *awakens* might not be the right word, for he has been not asleep but dead. Twelve days dead. But now Er is back among the living, and he has tales to tell.

The myth of Er comes at the very end of Plato's *Republic*. Throughout the *Republic*, Plato had been addressing the moral components of life, especially justice. Now, however, he was leaving those worldly — and Earth-bound — considerations behind. He wanted to turn his readers' attention to the *after*life. After the demise of his body, Er explains, his soul survived. It found itself joining a multitude of other souls on a journey that ended in a mysterious landscape punctuated by four holes, two in the earth and two in the sky, where judges awaited the newly arriving souls and evaluated each one. Those souls the judges pronounced "just" were to "continue their journey to the right and upward" — through one of the holes in the sky. There they would bask in a thousand years of "inconceivable beauty" before returning, "pure," to the present spot. Those souls the judges pronounced "unjust" were to "continue their journey to the left and down" — through one of the holes in the Earth. There they would endure a thousand years of suffering before returning "full of dirt and dust." Only then could the just and the unjust alike proceed to an eternity in the Elysian Fields.* The judges informed Er, however, that they had a special sentence for him: He was to observe the afterlife and report back to the mortals still living on Earth.

The fable had, as in the myth of Icarus, a moral and a meaning, which Plato would soon make explicit. But even at this early point in his narrative, Plato had already surprised his readers by including an option not generally available in previous mythology: upward mobility. And Plato was able to make this adjustment because he was in possession of new information about the realm that a soul might be upwardly mobile toward: what that realm actually was.

Plato often advocated in his writings for the study of astronomy, and when he founded his Academy in Athens, he made sure the subject was part of the philosophical program. Eudoxus of Cnidus, one of Plato's colleagues, realized that he could begin to approximate the motions *up there*—Moon, wandering stars (or *planētēs*), Sun, fixed stars—through math *down here* if he imagined celestial objects moving in concentric spheres.

And so, near the end of the tale of Er, the soldier encounters "a straight light like a pillar, most nearly resembling the rainbow, but brighter and purer." This light stretches from the Earth far into

---

* Plato paused here to acknowledge that even after a thousand years some souls required further cleansing. In their cases the "mouth" leading to the Elysian Fields "roared when one of those whose badness is incurable or who had not paid a sufficient penalty attempted to go up."

the heavens. In the midst of this pillar of light sits a woman named Necessity, and in her lap she holds a spindle. Extending from the whorl — the part of the spindle that spins threads outward — is a series of nested, umbrella-like bowls: the spheres ferrying the Moon, Sun, planets, and starry firmament in their orbits. The spindle of Necessity controls the circular motions of the major constituents of the cosmos because — as the name Necessity suggests — something must.

Plato didn't invent the idea of an afterlife. But before the myth of Er, the geography of the land of the dead usually imposed the same travel restrictions as the geography of the land of the living, as if even our immortal remains were still us: human, if different. The earliest record of the afterlife is six thousand years old. It tells the story of the Sumerian goddess Inanna, or Ishtar, as she descended into the Great Below, a relatively uneventful place to spend one's afterlife. No: after*lives* — a dual existence. The Egyptian body, or *ka*, stayed in its burial place at all times, but while the sun was out, the spirit, or *ba*, got to wander. During the day the *ba* might visit the world of the living or even travel with Ra across the sky; at night it would return to the burial place and reunite with the *ka*.*

A *ba* was fortunate. At least it could leave the netherworld. Not so the dead in most other cultures. They went below ground, and below ground they stayed.

---

* The contemporary mistranslation of the title of a collection of Egyptian funerary texts spanning a period of fifteen hundred years, *The Book of the Dead*, exactly reverses the original emphasis. To ancient Egyptians, it was *The Book of Coming Forth by Day*.

The place they inhabited tended to be numbingly nondescript at best, drearily mournful at worst. "In the realm of the dead," said the book of Ecclesiastes, "there is neither working nor planning nor knowledge nor wisdom." Job anticipated a "land of darkness and deep shadow."* No wonder that when Homer's Odysseus tried to reassure the ghost of Achilles that being dead can't be all that bad, Achilles snapped back, "I would rather be a paid servant in a poor man's house and be above ground than king of kings among the dead."

Exceptions to this middling existence might exist; truly extreme actions could call for truly extreme consequences. Greeks who challenged or defied the gods were banished to Tartarus, that lowermost region of the universe, as far down as *down there* goes. Their punishment wasn't just going nowhere. It was *getting* nowhere: Sisyphus, eternally pushing his boulder; the daughters of Danaus, endlessly attempting to fill leaky buckets; Tantalus, trapped in water, never able to raise his hands high enough to reach the breeze-nudged grapes.

Even so, if being *down here* was part of the human condition, then being *up there* must be part of the *super*human condition. And to be superhuman meant to be godlike.

"Behold," God said to Moses, when Moses visited him on Mount Sinai, "I am coming to you in a thick cloud, that the people may hear when I speak with you, and may also believe you forever." Three days later, God kept his promise. "Then the Lord came down

---

* To be fair, Job had good reason not to be an optimist.

upon Mount Sinai, on the top of the mountain. And the Lord called Moses to the top of the mountain, and Moses went up."

Moses came back down. The Lord, however, stayed up, and there he would remain. If we wanted to meet him, we'd have to do so on his terms. And the children of Israel did attempt to do so. Once they had established a presence in Jerusalem, several hundred years later, they chose the highest point in the city to place the Holy of Holies: a vast chamber, empty except for the spirit of God, which through this portal entered the Temple before whispering its presence into the world.

Jesus, too, encountered God on high. He led three disciples "up a mountain by themselves. And he was transfigured before them, and his face shone like the sun, and his clothes became white as light. And behold, there appeared to them Moses and Elijah, talking with him." Then Jesus, too, came back down.

Similarly, Muhammad: Up Mount Hira he climbed on an annual pilgrimage, there to isolate himself in a cave and spend weeks in prayer. His solitude ended the day the angel Gabriel appeared to him and inspired him to recite — as would happen on future pilgrimages as well — passages that would become part of the Qur'an. Technically, Gabriel wasn't God, but during that first visitation he did carry a message from a higher power: Muhammad is "The Prophet of Allah." And then Muhammad came back down.

Returning to Earth was the rule for humans. If you were a storyteller, however, and you wanted to confer a singular status on a character — if you wanted to confer not godlike status but the status of a god — you made an exception.

The prophet Elijah was out for a walk with his wife when "suddenly a chariot of fire appeared with horses of fire, and separated the two of them; and Elijah went up by a whirlwind into heaven."

Herakles was burning on his funeral pyre when the god Jupiter interceded. Jupiter directed the other gods to observe what would happen next. "He, who has defeated all things," Jupiter says of Herakles, "will defeat the fires you see." Not Herakles's body; that will burn. But the essence of Herakles will defeat the fires. It will be "immortal, deathless, and eternal: and that, no flame can destroy. When it is done with the earth, I will accept it into the celestial regions." Herakles, unlike all other Greek mortals except one, would become a god. The other exception was Ganymede, the boy whose beauty was "loveliest born of the race of mortals." Zeus, smitten, abducted him from the field where he was tending his flock of sheep and spirited him into the sky, there to spend eternity as the gods' cupbearer. Ganymede and Herakles achieved what no mountain-stacking whippersnappers ever could: entry into Olympus.

Jesus, too, rose from the dead, though not before raising Lazarus from the dead. But that's where the similarities between the two miracles ended. Lazarus eventually had to reverse course and retreat to his tomb, while Jesus left his tomb, rose, and just kept rising. Out for a walk with his disciples, "he was taken up into a cloud while they were watching, and they could no longer see him."

And that was that. A very few privileged mortals could achieve superhuman status in the sky. The rest of us had to resign ourselves to a hard fact of life: You didn't get a prize just for having lived — at least not until Plato came along.

The myth of Er was just one of Plato's fables of aspiration, and it was also just one of his fables of aspiration that relied on a *down here/up there* dichotomy. Elsewhere in the *Republic*, Socrates tells his parable of the cave, in which a freed prisoner who climbs upward after a lifetime in an underground lair discovers "reality" — our world. As in many other fables of Plato's, the speaker, standing in for Plato, is his teacher Socrates, and the student, standing in for us, is Plato's brother Glaucon. "You will not misapprehend me," Socrates tells Glaucon, "if you interpret the journey upwards to be the ascent of the soul into the intellectual world." In Plato's *Symposium* Socrates recounts a conversation with Diotima, a priestess from the Greek city of Mantinea, who proposes a hierarchy of learning — an "ascent to Beauty itself" during which you would be "ever mounting the heavenly ladder." She calls it the Ladder of Love: It begins, on the "lowest rung," with a physical attraction to one body; then climbs to the love of other bodies, of souls that transcend bodies, of laws and institutions and learning; and culminates on the uppermost rung, from which you would apprehend not a beautiful body nor a beautiful object nor a beautiful idea but Beauty itself: "one single form of knowledge."

The myth of Er also ended with a summation from Socrates for the benefit of Glaucon: The "worse" life is the one in which you become "more unjust," and the "better" life is the one in which you become "juster." But unlike Plato's other parables, the myth of Er was not merely aspirational. It was moral. It was about the improvement of the self with a purpose beyond the self: Seek knowledge, yes — but then use that knowledge in the cause of justice.

And in that respect, the myth of Er was more than a departure from Plato's other parables of aspiration. It was a break with the past. Before Plato, myths of the afterlife were mostly tales of passive predestination. How you would spend eternity depended on your station in life or on the circumstances of your death. But Plato was suggesting something radically different: How you might spend eternity depended on *you*. Your fate — at least the first thousand years of it — was determined not by who you knew but by how you behaved. To get *up there* you didn't need to perform a daring physical feat involving logs or wings; you needed to perform a daring feat of morality: the pursuit of justice for all. You still might not get a prize for having lived. But after Plato, you did get a prize for *how* you'd lived.

As for the spindle: It wasn't a prize, exactly. But it wasn't *not* a prize. It was, instead, a hint of a prize — a foreshadowing of what *up there* was about to become in the history of mythology: a heavenly reward.

Whether Plato believed in an afterlife is debatable. Possibly it was no more real to him than a soldier surviving a funeral pyre. And whether the West needed a Plato to formalize the moral divide between *down here* and *up there* is doubtful. The realm of the unapproachable and mysterious was a natural repository for a fantasy that we might have eternal happiness, while the realm of the inescapable and mundane — from the Latin *mundānus*, or "of the earth" — was a natural symbol of our resignation that we might not.

But Plato's specific landscape of rising and falling souls was new. It wound up dominating—and defining—the afterlife in the Western imagination. By the time Jesus began proselytizing on the shores of the river Jordan, some three and a half centuries after Plato wrote the *Republic*, the motifs of Er had suffused first Greek and then, after the sack of Corinth, Roman culture. A commoner such as Jesus, a carpenter from Nazareth, might not have heard of Er, but the lessons he preached, and the afterlife he described, were Plato's.

No longer was the afterlife a murky twilight where all souls, save a select few, coexisted in gloomy eternity. As in the myth of Er, it now consisted of two tiers. One was a place of reward—the "Kingdom of God," the description that Jesus used on multiple occasions. The other tier, the one below ground, was a concept still in transition.

The Greek versions of extreme punishment had been predominantly psychological. Sisyphus pushing his unraisable rock, the daughters of Danaus carrying their bottomless buckets, Tantalus reaching for his ungraspable grapes—these exertions were exercises in frustration, torments of the mind. So, too, were some of Jesus's descriptions of punishment: You will "see Abraham and Isaac and Jacob and all the prophets in the kingdom of God, but yourselves being thrown out." But Jesus added physical punishments as well—torments of the flesh taking place in a subterranean torture chamber. Jesus promises that God will say to sinners, "Depart from me, accursed ones, into the eternal fire which has been prepared for the devil and his angels."

Also as in the myth of Er, each of the two tiers was potentially accessible to all. Even the criteria for where one would spend eternity were similarly Platonic. "It is easier for a camel to go through the eye of a needle," Jesus tells his disciples, "than for a rich man to enter the kingdom of God." The disciples are agog: The rich man's fate is sealed just because he's rich? No, Jesus answers. The rich man's fate is sealed because he has *remained* rich. He's destined for eternal damnation not because of who he is but because of how he has behaved — or how he hasn't behaved: He hasn't redistributed his wealth among the less fortunate. You might call such a person unjust. And as for just — what could be more so than spending your life obeying the succinct admonition "Do unto others as you would have them do unto you"?

But echoing Plato, deliberately or not, also presented a problem. In the first couple of centuries after the death of Jesus, writings began to circulate that commemorated his philosophy: four short biographies of Jesus; an account of the acts that his disciples undertook to build the foundation of a new religion; epistles from those same apostles to potential converts; and a book of prophecies. Early Christian scholars considered these collective works a worthy continuation of the Hebrew Bible — a new covenant, a new testament. But they also thought these writings revealed that Plato's teachings on morality, while philosophically sound, were incomplete.

Plato had created a ladder of love that led to beauty. What Christian scholars needed was a ladder of love that led to salvation.

Four hundred years after the birth of Jesus — the number of years before or after Jesus's birth having become an increasingly common measure of history — Christian scholars got it: a ladder to a heavenly reward. In his *Confessions,* Augustine, a bishop of the African city of Hippo, described a series of rungs: the rejection of our innate love of temporal things, then a piety that "leaves no option" but to recognize the ultimate authority of scripture, and so on to the seventh rung, where true wisdom awaits. This thought completed the synthesis of Platonism and Christianity, but it also drastically changed the terms of the prize you might receive for how you lived. Plato's thousand-year separation of the just and the unjust ended in a reunion in the Elysian Fields. Augustine's separation was final: You went *up there* and stayed up there, or you went *down there* and stayed down there.

Metaphorical ladders were all well and good, but as the details of the Christian afterlife became more literal, so did the ladders in the collective imagination of the faithful. Less than two hundred years after the death of Jesus, a noblewoman named Vibia Perpetua, while awaiting martyrdom for her Christian beliefs in Carthage, recorded a vision: "I saw a ladder of tremendous height made of bronze, reaching all the way to the heavens, but it was so narrow that only one person could climb up at a time." The future saint Balthild, shortly before her death in 680, saw angels on ladders. The future saints Romuald of Italy and King Olaf of Norway, both of whom lived at the turn of the first millennium, saw monks. In 1348 the future saint Bernard Ptolemy saw both: angels leading white-robed monks.

In some ways *up there* hadn't changed much since Plato's tale of Er, and neither had his narrative motifs: Journey's end finds the hero contemplating the entirety of the workings of the cosmos from a perspective that reinforces the distance — geographical and psychological — between the majesty of the celestial realm and the modesty of the clot of dirt at its center.

In *Somnium Scipionis,* or *Scipio's Dream,* the first-century B.C.E.

Roman philosopher Cicero modeled his hero's journey on Er's, but Cicero chose as his hero a real historical figure: Scipio Aemilianos, the Roman commander at the destruction of Carthage, in 146 B.C.E., that led to the Roman Empire's acquisition of the Carthaginian Empire. Two years before that conquest, in Cicero's tale, the fictional version of Scipio has a dream. In it, Scipio's grandfather through adoption, Publius Cornelius Scipio Africanus, himself the general who defeated Hannibal at Carthage in 202 B.C.E., foretells the military glories awaiting the younger Scipio as well. As his reward he will ascend to join his grandfather and other great statesmen in the place "which you call," Scipio the elder tells Scipio the younger in the dream, "as you learned from the Greeks, the 'Milky Way,' and which you now can see" — for there the two Scipios are, floating among the spheres that carry the stars, the planets, the Sun, the Moon. At the center of the spheres rests the Earth, but its size bothers Scipio the younger: "Indeed so small did the Earth seem to me that I was ashamed of our empire through which we touch, as it were, merely a point of it."

"How long will your mind be fixed upon the ground?" the elder Scipio interrupts. "Do you not see what temple you have entered?"

He does. The young Scipio even hears its magnificence — the music of the spheres as they rub against one another, wheel greasing wheel. Yet Scipio, simple mortal, can't stop himself from turning his eyes back toward Earth.

Christian apocalyptic visions would continue the Greek and Roman tradition of a voyage to the farthest reaches of the universe, with one significant difference. The puniness of the Earth wouldn't be of secondary importance to the majesty of the heavens. It would be the point. In Saint Paul's Apocalypse, from the third century C.E., the guiding angel doesn't urge the narrator to bask in the glories of God. Instead, he says, "Look down on the earth." Paul obeys. "It was as if it were nothing to my eyes," he reports. "I saw the children of humanity as though they were nothing and utterly failing." Paul starts to look away, but the angel repeats the order. "The angel said to me look again upon the earth. And I looked and saw the whole world," Paul says. "Men and women were like nothing and utterly failing."

If the Christian journeys to the celestial realm were reminders of humanity's weaknesses, the Christian journeys below ground were . . . well, more of the same. But lacking a real-life counterpart to planets and the spheres keeping them aloft and on the move, visions of the underworld had to acquire a vividness of their own.

Part of the Christian canon that had become the New Testament was an Apocalypse, a type of text whose name means "revelation" in Greek. The genre had been around for more than a thousand years, but in the Christian era the purpose of Apocalyptic literature was most of all to frighten: Be a just person in this life so that *this* doesn't happen to you in the next.

The idea that the punishment should fit the crime — that the torture should correspond to the sin — dates at least from Saint Peter's Apocalypse, a second-century text that recounts what Jesus revealed about the End of Days to the apostle Peter just before ascending into heaven: "Those who have blasphemed the way of righteousness will be hung up by their tongues"; women who "plaited their hair, not for the sake of beauty but to turn men to fornication," will be hung "by their neck and by their hair"; "and the men who laid with them in fornication will be hung by their loins in that place of fire."

The tradition of the judgment happening at the moment of death dates at least from the Apocalypse of Paul, from the third or fourth century. "I would like to see the souls of the righteous and the sinners as they depart from the world," Paul says to the angel guiding him through space. The angel points him toward Earth just as someone there is dying. "Then," Paul writes, "I heard a voice saying, 'Let that soul be delivered into the hands of Tartaruchus'"—the traditional guard of Tartarus—"and he must be taken down into hell.'"

In Tundale's Vision, from 1149, the punishment for fornicators was pregnancy-by-beast, then birthing-of-beast—"not only women, but also men. Not through the part that nature constructed suitable for such a function, but through their arms, just as through their breasts, and they were bursting out through all members." Only four years later, in Saint Patrick's Purgatory, the Knight Owen observed the victims of an impressive array of punishments: "Some were suspended over fires of brimstone by iron chains fastened to their feet and legs, with their heads downward; others hung by their hands and arms, and some by the hair of their heads. Some were hung over the flames by hot iron hooks passed through their eyes and noses, others by their ears and mouth, others by their breasts and genitals." In 1196 or 1197, the Monk of Esham described his vision of the damned: "The torturers ran to them with forks, torches, and every sort of instrument of torture and returned them back to their punishment again to inflict every kind of cruelty on them." Such as? Would you perhaps be willing to elabor— "Some were roasted before fire; others were fried in pans; red-hot nails were driven into some to bare bones; others were tortured with a horrid stench in baths of pitch and sulfur mixed with molten lead, brass and other kinds of metal; immense worms with poisonous teeth gnawed some; others were fastened one by one on stakes with fiery thorns."

A hundred years or so later, in the first decade of the four-teenth century, the Italian poet Dante Alighieri set out to write his own contribution to the genre. Dante modeled his *Commedia** on the journey-through-the-afterlife sections in the two great epics of Western literature, Homer's *Odyssey* and Virgil's *Aeneid*. Just as Virgil nodded to Homer, so Dante-the-author would nod to Virgil, assigning him a role as Dante-the-character's guide for part of his explorations. Homer, writing in Greek, had defined the afterlife for Greek literature. Virgil, writing in Latin, had defined it for Roman literature. In the thirteen hundred years since Virgil, the pagan civilization had fallen and a new civilization had arisen, and that civilization had produced its own narrative tropes. Dante, writing in Italian, would define the afterlife according to that model. Dante's afterlife would be the afterlife of Christianity.

Like his readers, Dante believed the afterlife to be as literal as a ladder, and his epic would adhere to that belief. His epic would be literary as well; he would take poetic license. But even though his afterlife was fictional, its residents and their realms were not. To Dante and his readers, heaven and hell — or Heaven and Hell, as they thought of them — were real. Angels and demons existed. God and Lucifer existed. Eternal rewards and eternal punishments existed.

His journey would begin, in the *Inferno,* with a descent *down*

---

* This was Dante's title. *La Divina Commedia,* or *The Divine Comedy,* is a later imposition.

here, and it would end, in the *Paradiso*, with an ascent *up there*, and because he was writing about the Christian afterlife, he would have to venture farther in both directions than any of his predecessors.*

The journey *up there* would follow the now-familiar formula: Dante-the-author sends Dante-the-character on a tour of the same spheres that had been around since Plato, though because Dante was writing in the Christian tradition, he got to add a sphere higher than the rest: "Beyond all these the Catholics assert the empyrean heaven," as he had written several years earlier, in an eventually abandoned work he called *Convivio*, or *The Banquet*, an attempt to catalogue the whole of human knowledge. "They assert it to be immovable"; "still and tranquil is the place of that supreme deity which alone completely perceiveth itself." The epiphany in the *Paradiso*, however, was the same as the one he'd inherited: "I turned my gaze back through each and every one of [the] spheres, and saw this globe, so that I smiled at its pitiful semblance, and I approve that wisdom greatest which considers it least: since he whose thoughts are directed elsewhere may be called truly noble." The Earth was still puny, and its inhabitants were still pitiable.

The *Inferno*, though, gave Dante the opportunity to add his own variations on the Apocalyptic literature. Punishments fit the crime:

---

* Dante also goes farther *middle*, so to speak: In the middle volume of the *Commedia*, Virgil and Dante visit a mountain — Purgatory, a way station between Earth and Heaven, a conceit that had entered the Christian vocabulary only a hundred years or so before the birth of Dante.

Nine circles of Hell correspond to nine levels of sin. Judgments are final: "Abandon hope, all ye who enter here," reads the inscription at the entrance to Hell. As for vivid descriptions, take your pick: the corrupt nobleman forever feasting on the brain of an even more corrupt archbishop; popes plugged head first into rocks; Judas Iscariot dangling from the mouth of a perpetually masticating Lucifer.

But if Dante were to create the journey-through-the-afterlife narrative that plunges the reader farther down than any previous journey-through-the-afterlife narrative, he would have to confront a further challenge to his imagination. Amid the knowledge that Dante had included in his *Convivio* was a fact that he knew would now test his storytelling skills: The Earth is round.

Homer, writing eight hundred years before Christ, hadn't known that. Cicero, writing in the first century B.C.E., probably had; by then the Greeks had considered it common knowledge for several centuries. But Cicero hadn't pursued this facet of the underworld. Neither had the Christian Apocalyptics. Dante would, and in doing so he inserted into mythology something resembling what several centuries later would come to be called science.

At the end of the *Inferno*, Dante and Virgil encounter Lucifer, who is stuck in ice up to his chest. Hold on to my neck, Virgil instructs Dante, and together the two of them climb onto Lucifer's back. From there they descend "from tuft to tuft between the matted hair and the frozen crusts," but when they reach Lucifer's groin, Virgil pauses. Then he begins to contort himself. Through great effort Virgil reverses his vertical position until his head is now where his feet had been, and vice versa. Then he begins climbing again, and

Dante fears that they are descending back into Hell. No, Virgil reassures Dante. "When I turned myself," Virgil says, "thou didst pass the point to which weights are drawn from every part" — the center of the Earth, which is to say, the center of the universe.

Dante's characters have come out the other side, and so has mythology. By following the idea of a round Earth to its logical extreme, Dante has eased the symbolic hold of the horizon on the human imagination. He's guided his characters so far *down here* in their fictional universe that when they eventually emerge, they'll be heading not just toward Paradiso but in the direction they would be heading in the physical universe: *up there*.

2

GRAVITY IN MATTER

The conversation was ancient. It was already ancient when Aristotle joined it. "The ancients," Aristotle wrote, "gave to the Gods the heaven or upper place, as being alone immortal." Aristotle agreed: Ancient philosophers had gotten at least that much right in describing the celestial part of the cosmos. Not so with the terrestrial: There, Aristotle felt, they'd gotten almost everything wrong.

Before Aristotle, knowledge of the natural world was haphazard and incomplete and unreliable. Aristotle's teacher, Plato, had re-examined philosophy at such length and depth that his works had the effect of being a fresh start — and a comprehensive one at that. Aristotle assigned himself the same task in the natural world: not to fantasize about the workings of the universe but to discover how the universe actually works. To begin at the beginning, again.

"The inquiry into nature," Aristotle pronounced in *De Caelo,* or *On the Heavens,* from the mid-fourth century B.C.E., "is concerned with movement": the motions of matter. As was the case for the mythmakers and the storytellers of old, Aristotle needed to clear the landscape of everything except what was essential to the universe. As was also the case for the mythmakers and the storytellers, Aristotle as a result found himself confronting two realms: the knowable and the unknowable; the terrestrial and the celestial; the mundane and the mysterious.

In the new beginning? The old dichotomy: *Down here. Up there.*

The sky hadn't fallen. The ancients knew that much for sure. What they couldn't know was whether whatever was up there would one day fall down or even could fall down. That's what things do — fall *down here,* eventually — and so, we might assume, the sky probably could and would.

The Greek poet Theognis of Megara called the possibility "that terror of earth-born men." When the Roman playwright Terence needed a metaphor for timidity, he could cite "the folk that say, 'What if the sky were to fall this very moment?'" When the Celts of the Adriatic sent envoys to create an alliance with Macedon, their host Alexander asked them whom they most feared. Not a who, they said. A what: the sky, falling on them.* In his *History of Rome,*

---

* Wrong answer! Right answer: "You, Alexander." They reached a treaty anyway, if only because Alexander could hardly fault them.

Livy recounted what happened to the Bastarnae when they pursued the Thracians into the Donuca mountains: "Not merely were they assailed with a deluge of rain and then masses of hail, along with tremendous crashes in the sky and thunders and lightning-flashes blinding their eyes, but also the bolts flashed all about them so that their own bodies seemed to be the targets, and not only the common soldiers but even the chieftains fell stricken on the ground." The Bastarnae retreated, only to cross paths with their ostensible prey, the Thracians; fortunately, they had a reasonable explanation for their retreat, one that didn't make them look like cowards: The sky was falling.

Presumably when the rains passed and the sun emerged and birds took wing once more, the Bastarnae would have joined the rest of the world in noticing that the sky *still* hadn't fallen. And the longer it continued not to fall, the more one might wonder why not. What's holding it up?

One Egyptian myth proposed that the Earth supports the sky, through four pillars that mark the cardinal directions, plus one central pillar. The Cherokee proposed the opposite relationship: It's the sky that sustains the Earth, which hangs from the heavens by four ropes. The Greeks split the difference: The flat Earth is inside a globe, or, if you prefer, a globe is outside the flat Earth; either way, the ends of the Earth define the diameter of the globe as much as the diameter of the globe defines the ends of the Earth. And that diameter, according to Hesiod's *Theogony,* is as follows: Drop an anvil from the uppermost reaches of the globe, and it will fall for ten days before hitting the Earth. Drop an anvil from the Earth, and it will fall for ten days before hitting the bottommost reaches of the globe.

Aristotle rejected them all—every "old tale which says that the world needs some Atlas to keep it safe." In trying to understand the rationale of the mythmakers, Aristotle identified two incorrect assumptions they had made right at the beginning.

The first affected the interpretation of *down here*—that the Earth is flat. Aristotle had determined for himself that it's not. He had observed a lunar eclipse—the singular phenomenon in which the Sun casts a shadow of the Earth on the Moon; the shape of that shadow is unmistakably a circle. Aristotle had noticed, too, that distant ships rise over the horizon, as if climbing a hill, and that if you travel sufficiently north or south, new constellations will emerge before you and familiar constellations will vanish behind you. No, he concluded, the Earth is round.

The second incorrect assumption affected the understanding of *up there*—that it consists of the same matter as *down here*. Mythmakers, he wrote, "conceived of all the upper bodies as earthy and endowed with weight." If that assumption were accurate, then yes, the Earth would need an Atlas or pillars or ropes to ensure its survival. But why assume that the Earth and the heavens consist of the same matter, especially when a little reasoning demonstrates that they don't?

Reasoning, in fact, was what Aristotle would introduce into the conversation: methodology, not mythology. Ancient storytellers had begun their ruminations on the motions of matter with sense evidence—what they could see—and then resorted to their wildest imaginations to explain the otherwise inexplicable. Aristotle, too, felt he had no choice but to start with sense evidence. But then, he vowed, he would apply only logic.

He began *down here*. The motions of matter with which the contemplators of nature who preceded Aristotle had concerned themselves was the rate at which different kinds of objects fall. Those that fall faster they deemed heavier; those that fall slower, lighter. Aristotle found that description to be obviously accurate. But he also found it to be incomplete.

The ancients were correct in observing that things fall. Clots of earth, of course, drop to Earth, and water collects in depressions in the Earth's surface or soaks into the ground.

But the ancients were incorrect in assuming the existence of only one direction of motion. Doesn't fire rise? Doesn't air bubble to the water's surface, then break it?

The inquiry into the motions of matter *down here*, Aristotle concluded, must therefore concern itself with not one movement — downward — but two movements, each equally legitimate: downward *and* upward.

Because the ancients' conception of motion went in only one direction, the most they could investigate was *heavier than* and *lighter than*. Aristotle provided an example. A bronze object is going to be heavier than a wooden object; it will fall faster. A wooden object is going to be lighter than a bronze object; it will fall more slowly. The bronze object is indeed going to possess more *gravity* than the wooden, and the wooden object will indeed possess more *levity* than the bronze.

But in Aristotle's system — a system in which objects move in one of two directions — comparisons of heaviness or lightness wouldn't always make sense. You can say that bronze falls faster than wood, but you can't say that bronze falls faster than fire, because fire doesn't fall. You can say that wood falls more slowly than bronze, but you can't say that wood rises more slowly than air, because wood doesn't rise. Relative comparisons work only when comparing like with like — only when employing, as Aristotle said, "the absolute use of the terms."

"By absolutely light," he went on, "we mean that which moves upward or to the extremity" — fire and air — "and by absolutely heavy that which moves downward or to the center" — earth and water.

His predecessors, Aristotle concluded, hadn't looked closely enough, hadn't thought hard enough. They hadn't appreciated the connection between looking and thinking, between evidence and logic.

In reaching these conclusions Aristotle had begun with sense evidence, phenomena he could see and touch. By examining that evidence, he had discovered two motions in nature — straight up and straight down. But if he wanted to continue his inquiry into the movements of nature, he would have to pursue where those movements led — "upward or to the extremity" and "downward or to the center." And in that case, he would have to rely less on the evidence of the senses and more on logic.

Defining the center of the universe was the easier of the two. Not only did Aristotle know that the Earth is round, but he could pair that idea with the fact that wherever on this round Earth things possessing absolute heaviness fall, they fall not just down but *straight* down. You can drop a clot of dirt here, then go one hundred or one thousand or ten thousand stadia in any direction and drop another clot of dirt, and both clots will fall in the same straight line. Both will intersect the locally flat surface of the globally round Earth at the same right angle — meaning that they must be falling toward a common center. You can impede that fall; you can, for instance, stop the fall of a clot of dirt with the palm of your hand. But once you remove the impediment, the clot of dirt will resume its natural, straight downward motion, at least unless it hits another impediment, which you would then remove, and another, which you would remove, and so on. At some point the clot of dirt will reach a final, impassable impediment, the one you can't remove: the surface of the Earth.

But what if you could remove the surface of the Earth? If it is in fact just one more impediment similar to the others, its removal will allow the clot of dirt to resume its straight downward descent toward a common center. That center is the center of the Earth. But, Aristotle wondered, what if two Earths existed? Would the objects on each Earth fall to their own center? In that case, the universe would have two centers — or more: as many centers as there are Earths. No, Aristotle reasoned, the center of the Earth is not the destination. It only *appears* to be the destination. The actual destination is the center of the universe, the true destination of all objects possessing absolute heaviness. Remove the Earth entirely and heavy objects would still fall toward the same center of the universe. If two Earths existed, they would both fall to the center of the universe, like the clots of dirt that they are. The center of the Earth and the center of the universe occupy the same location simply because everything heavy falls toward the center of the universe — at least the part of the universe below the heavens.

As for the heavens — the part of the universe toward which fire and air rise; the part of the universe that is "upward or to the extremity" — the evidence of the senses was even more sparse. "We have but little to go upon," Aristotle lamented, "and are placed at so great a distance from the facts in question." He couldn't actually examine the heavenly bodies at close range or manipulate them for himself, as he might do with fire or a clot of dirt. He would have to rely only on whatever little he could observe at this distance and apply an even greater helping of logic.

The syllogism was perhaps Aristotle's favorite rhetorical device, and while he didn't explicitly invoke that form in *De Caelo*, he implicitly followed it. He began with what he knew to be incontrovertibly true: "Our eyes tell us that the heavens revolve in a circle." Then he constructed his argument:

Objects that move in circles move neither up nor down.
Celestial objects move in circles.
Therefore: Celestial objects move neither up nor down.

And:

Objects that move neither up nor down are neither light nor heavy.
Celestial objects move neither up nor down.
Therefore: Celestial objects are neither light nor heavy.

And:

Objects that are neither light nor heavy must be indestructible and
unchanging.
Celestial objects are neither light nor heavy.
Therefore: Celestial objects must be indestructible and unchanging.

The ancients had a name for hypothetical matter that was as pure as Aristotle was claiming — matter indestructible and unchanging, a fifth element beyond the familiar four of earth, air, fire, and water: *aether.* If heavenly bodies moved along Eudoxus's spheres, Aristotle reasoned, then surely the spheres themselves were aethereal.

One more step remained in Aristotle's new methodology of applying logic to the motions of matter *up there*: corroboration. And he found it: observations of the heavens dating all the way back to the Babylonians and Egyptians. "For in the whole range of time past," Aristotle wrote, "so far as our inherited records reach, no change appears to have taken place either in the whole scheme of the outermost heaven or in any of its proper parts."

Aristotle had succeeded in his self-assigned mission: the re-creation of the heavens and the Earth. And then, on the seventh day — to coin a phrase — he rested.

The conversation was ancient again, only this time the ancients were Plato and Aristotle.

John Philoponus's regard for Aristotle had started high. As a young man in the early sixth century C.E., he belonged to a group of scholars in the Roman outpost of Alexandria, the Egyptian metropolis on the Mediterranean coast. He and his fellow students would gather in one of the lecture rooms, settling on the two sets of stone ledges that curved outward from a central *thronos,* or throne—a seat upon which, several steps up, perched their instructor, Ammonius Hermiae. Opposite the throne, in the space between the two sets of benches, stood a podium at which a student or Ammonius himself would lecture or answer questions or defend an argument. More often than not, the topic under discussion was how to interpret ancient texts.

John Philoponus—Philoponus being a nickname meaning "lover of toil," and John being shorthand for Christian—and his fellow students were doing what Greek philosophers within the Roman Empire had been doing for centuries: writing commentaries. These commentaries were respectful summaries and close analyses of the writings of the ancient philosophers, particularly those of Aristotle and Plato, texts that were now more than eight hundred years old. To Plato and Aristotle had fallen the enviable responsibility of explaining everything. To Platonic and Aristotelian scholars had fallen the responsibility, enviable or not, of explaining Plato's and Aristotle's work to themselves and to one another.

By 529 C.E., however, Philoponus was having doubts about the foundations of Western philosophy itself. That year he initiated a decade-long reconsideration. In his first tract, *De aeternitate mundi contra Proclum,* or *On the Eternity of the World Against Proclus* (who had been Ammonius's teacher), he mounted an indirect assault on Aristotle. In his next writings, beginning with *De aeternitate mundi contra Aristotelem,* or *On the Eternity of the World Against Aristotle,* he cut out the middleman.

In considering the matter in the heavens, Aristotle had begun with a logical assumption and concluded with the corroboration of sense evidence, neither of which Philoponus found convincing. First, Aristotle assumed that matter *up there* was different from *down here* only because it was mysterious — because it was impossible to investigate at close range. Its motions *appeared* to be different, so Aristotle assumed that they *were* different, and he argued from there. Then he went looking for corroborating evidence among ancient astronomical tables and found no change in the motions and positions of celestial objects. But what, Philoponus argued, was a period of one or two thousand years on the scale of history? Had Aristotle learned nothing from his own consideration of the shape of the Earth? Although Aristotle knew that the world is round, he understood why you might easily believe it's flat: because it certainly looks flat here, and here, and here, and here and here and here. Maybe the same reasoning applied to time. Maybe a thousand years is the temporal equivalent of a flat Earth — now, and now, and now, and now and now and now. Just as a lot of *heres* don't add up to all of Earth, maybe a lot of *nows* don't add up to all of history.

Even so, the possibility that the heavens might exist on a time scale long enough to allow alteration doesn't mean that they actually *do* exist on a time scale long enough to allow alteration. Philoponus had identified a flaw in Aristotle's logic, but a flaw suggested only that Aristotle *might be* wrong. What Philoponus needed was a convincing demonstration that Aristotle *was* wrong — and he needed it because it's what Christianity demanded.

The problem was all this "In the beginning" business. How the sky and earth came to be. Whether the universe always existed or came into existence. Whether matter was eternal or created.

To the traditional Greek mind, the idea that matter could come into existence was nonsensical, absurd. The *arrangement* of matter — sure. The contents of the cosmic egg, the Chaos of undifferentiated matter, the primordial water and the primordial slush and the primordial P'an Ku — the raw materials of the universe had always existed. At some point those raw materials diverged into sky and earth. That moment of separation is what nearly all cultures meant by "creation." It was what the Greeks meant by "creation."

It was not, however, what the Christians meant by "creation," because in the beginning there was nothing to separate. Creation in the Christian teaching came out of nothing — *ex nihilo* — because out of nothing is what the Christian reading of the Bible required: "In the beginning God created the heavens and the earth." There it is — first book, first page, first words. God didn't *shape* preexisting matter. God didn't take what was already there and breathe life into it and give it form and arrange its positions and set it in motion. God *created* matter — a proposition making Genesis an exception among creation stories: In the beginning God created the heavens and the earth, but before the beginning was only God.

In 313 C.E., the emperor Constantine — himself a recent convert to Christianity — oversaw the Edict of Milan, which required tolerance of his adopted religion. Since then, the once-persecuted Christians had become the new persecutors, plaguing the pagans in part because they rejected creation *ex nihilo*. Following a particularly brutal purge of rebellious pagans in 488–499, Philoponus's teacher Ammonius Hermiae was one of the few pagan philosophers remaining in Alexandria. While he continued to teach well into the sixth century — he died at age eighty, in 520 — his students merely tolerated the part of his pedagogy that included his advocacy of a beginningless universe; one of them composed a satire in which Ammonius would repeatedly parry with his students over the subject, and invariably lose.

Still, Christian philosophers found themselves somewhat on the defensive regarding creation *ex nihilo*. The problem wasn't that they lacked an authoritative source, since "The Bible says so" was justification enough. What their argument lacked was a supporting logic.

Philoponus provided it — and he found it by using Aristotle's own logic *contra* Aristotle.

Among Aristotle's innumerable philosophical achievements had been his reconsideration of the mathematical concept of infinity. Prior to Aristotle philosophers had assumed that an infinite number or quantity would be a number or quantity that encompasses all possibilities — a number with no number beyond it. But Aristotle argued that every number has numbers beyond it — numbers you can add on, as Aristotle wrote, "ever other and other." Like the horizon on a round Earth, it provides a direction you can head in, a target you can keep going toward, even if you can't ever reach it.

Philoponus, however, realized that in terms of the passage of time the horizon of infinity lies in only one direction. Moving forward in time, the concept of infinity follows Aristotle's principle: a finite number to which you can always add another number — another day, another year. Whatever the current number of days or years is right now, at some later point in time — 24 hours later or 365 days later — it will be the current number of days or years plus one, producing a new current number, and so on, ever other and other. You can indeed just keep going in that direction and never stop.

Going in the other direction, however, challenged Aristotle's conception of infinity. Without a beginning, Philoponus argued, the measure of time has no moment to start counting from. And without a moment to start counting from, the measure of time requires not a number of days or years starting at a certain *now* and then stretching forever, but a number of days or years *not* starting at a certain *now* and then stretching forever. Not a finite number to which you can add another number, but an infinite number to which you can add another number — an infinite number of days or years, plus one.

And infinity plus one, Philoponus continued, is a number greater than infinity. It's a number beyond which no number can exist . . . plus a number.

For Philoponus, a number greater than infinity would be illogical enough. But a beginningless universe also requires *unequal* infinities. The number of times the Sun and the planet Jupiter have circled Earth, for example, would each be infinity, but because the Sun circles Earth once every day while Jupiter circles Earth over the course of years, the Sun's infinity would be many times greater than Jupiter's infinity.

No, Philoponus concluded, logic dictates that the universe had a beginning. Not a pseudo-beginning before which existed a formless substance or substances, but a proper beginning — a beginning before which nothing existed. A beginning at which everything came into existence. *Everything* — terrestrial and celestial, *down here* and *up there*. And an *up there* that had a beginning is a celestial realm that is capable of generation or destruction — and therefore is not composed of the perfect fifth element, aether. It's just more of the same old four.

De aeternitate mundi contra Philoponum was, perhaps inevitably, the title that Philoponus's most fervent critic contributed to the discussion. Simplicius of Cilicia* would have had good reason to begrudge his contemporary even if he'd respected Philoponus. To Simplicius had fallen the misfortune of being a pagan philosopher working at the reincarnation of Plato's Academy in Athens when — also in 529 — the emperor Justinian closed it and decreed that philosophers could no longer teach the pagan interpretation of creation (though they could continue to write about it). Simplicius felt that Philoponus, a Christian in Alexandria, continued to prosper for no reason other than that his political leanings were in fashion. Retreating from Athens to the hinterlands in exile, Simplicius read Philoponus's *Against Aristotle*, and, seething, made it his life's mission to expose Philoponus as a fraud.

Like Philoponus, Simplicius was born around 490; also like Philoponus, he studied under Ammonius in Alexandria. But the two nascent philosophers hadn't concurrently occupied the benches before Ammonius's *thronos*. Philosophy had not been Philoponus's first stop in academe. Before he studied under Ammonius, Philoponus had been a grammarian — a professional false start that Simplicius would never tire of mentioning.

"The Grammarian," Simplicius called Philoponus in his writings, or, even more dismissively, "this guy." Simplicius considered

---

* A Roman province along the southern, Mediterranean border of modern-day Turkey.

Philoponus to be inferior in erudition, culture, morality. Philoponus, Simplicius argued, had some mental deficiency that affected his ability to reason. Or he was a drunk. Or a madman. Or all of the above. "I have tumbled into the Augean stables," wrote Simplicius, for Philoponus had amassed "a bed full of dung." Worst of all, Philoponus was blasphemous.

At a different point in history, Simplicius felt, Philoponus's writing might not have merited his attention. Regarding the logic behind Philoponus's rejoinders to Aristotle, Simplicius is dandruff-dusting dismissive: "His objections are beside the point." But *Against Aristotle* itself demanded a response because of what it represented: the spiritual bankruptcy behind Christianity.

Simplicius was clinging to a hope, common among pagan scholars, that Christianity was a blip in history, more fad than philosophy. Their objection wasn't just that the creation of matter out of nothing was ridiculous — though ridicule it they did. It was that creation *ex nihilo* was irreligious. The Christians might cast it as a spiritual imperative, what with God bringing forth the entirety of everything in a singular display of omnipotence. But at what cost to religion itself? A celestial realm not composed of aether is a celestial realm that is literally unexceptional.

"This Grammarian," Simplicius wrote in his *Against Philoponus*, "regarded it as a matter of great importance if he could entice large numbers of laymen to disparage the heavens and the whole world as things that are just as perishable as themselves, and evidently to disparage the world's divine craftsman" — the god or gods or, in Plato's vague word, demiurge that had created order out of chaos. "For the whole intention of these [Christian] people's piety is to show that both the heavens and the heavens' creator do not differ from them in any respect."

You could see evidence of their arrogance in the belief that their God had been a man (even if, so the story went, his body was spared the indignities of the grave by ascending into heaven). You could see it in their faith that the relics of martyrs — corruptible, imperfect, impermanent substances: flesh, bone, cloth — were as worthy of veneration as the incorruptible, perfect, permanent heavens. You could see it in their belief that their souls might ascend there, as if they themselves were gods.

Too late. The discussion was over even as Simplicius tried to reframe it: Philoponus had persuasively answered the pagan arguments by turning their Aristotelian logic to his own advantage; the emperor Justinian had enshrined Christianity as the sole spiritual subject capable of public discussion. Simplicius's legacy, apart from his writings, would be a case study in *wrong place, wrong time.*

Philoponus's legacy had its own problems. About a century after his death, the Eastern Orthodox Church condemned him as a heretic for his interpretation of the Trinity as three separate deities rather than as one God in three manifestations. As Simplicius had hoped, numerous portions of Philoponus's manuscripts as well as entire tracts went missing over the centuries. But they weren't lost to posterity. Thanks in part to all that work by Simplicius to summarize and quote at length from Philoponus's writings, the arguments of the Grammarian survived, and as long as they were in the literature, their revival remained a possibility.

◈

The conversation was ancient yet again, only this time the ancient philosopher was Philoponus. In the early 1590s, the aspiring Dutch astronomer Johannes Kepler imagined himself on the far side of the terrestrial–celestial divide, saw what he could see, and reported back.

Kepler had been observing the Moon, and he had concluded that the gradations in shade were very likely to be mountains and valleys and plains. He imagined a landscape — a Moonscape — similar in some significant respects to his own Earthscape. He had no means of verifying his hunch; like all observers throughout human history, he had, in Aristotle's words, "but little to go upon." All he had were his eyes, and they were hardly adequate for distinguishing much more than those gradations of gray — the same gradations seen by generation after generation after generation of sky-gazers, who had concluded, after exercising various logical exertions, that such seeming irregularities were no threat to Aristotle's aether. But Kepler had his own logical exertions, and he was able to imagine not only a Moonscape with mountains but the view from its valleys — a unification of *down here* and *up there* that had recently graduated from philosophical possibility to physical probability.

The collection of wisdom about the natural world that the philosophers of Kepler's era had inherited was in its own way just as haphazard and incomplete as the wisdom that Aristotle had inherited and just as unreliable as the wisdom that Philoponus had inherited. Knowledge, of course, is never monolithic. Within any era not everyone agrees on what the current body of knowledge means or even on what the current body of knowledge is, and not all the knowledge available in one era passes to the next generation, whether because of the unavailability of the sources (texts go out of circulation) or the indifference of the interpreters (schools of thought go out of fashion). And of the information that does pass to the next generation, not all passes equally — equally in importance, in completeness, in comprehension. When scholars of a certain era assert that scholars of an earlier era held certain beliefs or knew certain information, what they mean is that they know from the historical record that certain beliefs or knowledge were under widespread discussion and seem to have reached some level of consensus.

These caveats apply especially to the period in the Christian West between the time of Philoponus and the time when his works might have a second life. When Muslims invaded Europe in the seventh and eighth centuries, they found only the husk of Greek scholarship — documents — and not the activity of scholarship itself. The Muslims preserved the manuscripts, made translations of their own into Arabic, and often added their own thoughts and discoveries; by the tenth and eleventh centuries, those Arabic versions were making their way back into Christian circulation, where they underwent translation into Latin and became the basis of scholarship well into the Renaissance.

In the case of Philoponus, his arguments reached the Islamic scholars Avicenna and Abu'l-Barakāt al-Baghdādī, whose writings about Philoponus kept his ideas circulating for the next few hundred years. Aristotle's spheres were ascendant again, because Aristotle was ascendant again, but so was the Christian doctrine of creation *ex nihilo*, which required a non-separation of *down here* and *up there*. So uncertain was the state of scholarship that you could hold conflicting views, secure in the knowledge that you could call upon one or the other in accordance with the needs of the moment. Explain the motions of the heavens? Try Aristotle. Interpret the doctrines of Christianity? Open a Bible. Besides, just because the Aristotelian system required spheres to keep the heavenly bodies turning didn't mean those spheres were necessarily made of aether. Similarly, just because the Christian system required *down here* and *up there* to consist of the same material didn't mean spheres couldn't exist. The current state of knowledge might have seemed vague, but that was only because it *was* vague, at least until more knowledge came along.

And then: More knowledge came along. Most of that knowledge scholars could find in translations of ancient manuscripts — not just Aristotle's but those of other philosophers whose works scholars were rediscovering on a regular basis. For Philoponus, the revival began with the first republication of the original Greek version of *Against Proclus* in 1535; Christian scholars praised Philoponus's arguments in favor of creation *ex nihilo*. Only eight years later, the Polish mathematician and cleric Nicolaus Copernicus published his Sun-centered system of the universe, *De revolutionibus orbium coelestium*, or *On the Revolutions of the Heavenly Spheres*.

Copernicus had written *On the Revolutions* partly in response to a direct request from the Holy Roman Church to come up with a mathematical system that was more consistent with the motions we see in the heavens than the current travesty. The question was the same one that Aristotle had asked: What's *really* going on up there? Aristotle had relied on logic for his answers, but several motions defied logic — for instance, retrograde motion, the period when the steady night-by-night, west-to-east progression of a planet stops and reverses itself for a few days before righting itself and resuming its previous eastward course.* A spheres-only cosmology couldn't explain that behavior, but, as philosophers came to discover, a spheres-and-then-some imaginarium could — circles, and circles within circles, whirling in orbits tangential to the spheres. At that point in the argument, however, the purpose shifted. No longer was it to capture what was really happening by applying logic; instead, it was to "save the phenomena" by appending geometrical constructs. It was to force the math *down here* to match the motions *up there*.

The Greco-Roman mathematician Ptolemy, writing circa 150 C.E., added some geometric contributions of his own, then formalized the whole system in a single volume that, nearly a thousand years later, Arabic scholars would deem *Al-mageste,* or *The Majestic One,* a title that European scholars later condensed to *Almagest.* By the sixteenth century, however, the math in the *Almagest* no longer matched the motions; the geometrical approximations had compounded their errors over time until seasons strayed and holy days drifted. The mathematical representation of the motions of the universe was an assortment of desperate measures that, as Copernicus said in his introduction to *On the Revolutions,* resembled "a monster rather than a man."

The Church asked Copernicus to try to solve the problem, and to a large extent he did. His solution was to move the center of the universe from the Earth to the Sun. Such a strategy certainly had the

---

* Today we know that the seeming irregularity of the motions is due to an optical illusion: The orbits of Earth and another planet will "overtake" each other because the planets are traveling at different speeds.

advantage of supporting the *down here/up there* sameness required by creation *ex nihilo*, but it had the disadvantage of overthrowing just about every other assumption in cosmology. Copernicus withheld publication of *On the Revolutions* until the end of his life; a copy reached him in 1543, when he was seventy and on his deathbed. He needn't have worried. If you wanted — and religious leaders did want — you could interpret his book not as a referendum on what was *really* up there but as a better way to save the phenomena in an Earth-centered cosmos, even one full of invisible (but not aethereal!) spheres.

In 1572, however, the world bore witness to the seemingly impossible: a new star. It appeared there one night, out of nowhere. Could the unalterable heavens really undergo alteration? Yes, at least if the spheres were more metaphorical than real. They had better be metaphorical, anyway, if they had any chance of surviving in the imagination of philosophers. Five years later, in 1577, another seeming impossibility under a strictly Aristotelian cosmology appeared: a comet that lit up the sky for weeks.

The historical record was full of comets and the sudden appearances of starlike objects. In the Aristotelian interpretation these disturbances were the result of an overabundance of fire that had followed its natural path straight up from the Earth to the first, nearest sphere — the one containing the Moon — and then, having nowhere else to go, had dispersed in an exhibition of light visible from Earth.

The night sky, however, had never been studied by an observer with the talents of the Danish astronomer Tycho Brahe. So renowned was Tycho that the Danish king granted him an estate on the island of Hven along with the resources to build the world's greatest observatory, Uraniborg, complete with instruments allow-ing him to make the most accurate astronomical measurements in history. The new starlike object of 1572 didn't change position even over the course of months. Tycho concluded that it must lie among the fixed stars—indeed, as he argued in his *De nova stella,* or *On the New Star,* it *was* a fixed star. His argument regarding the comet was even more persuasive; he could calculate its trajectory, so he knew that the comet was hurtling through the celestial realm, not the sublunar realm, and he knew that it would be cracking sphere after sphere among the planets, if they were really there in a physi-cal sense.

The nonexistence of physical spheres—even the nonexis-tence of metaphorical spheres—wasn't a decisive argument against the Aristotelian/Ptolemaic system. From the start the spheres had been, in a sense, man-made; philosophers had appended them to the physical system of the heavens to help make sense of the mo-tions. Take away the physical spheres, make them as metaphorical as any bear or archer or water-bearer among the constellations, and the motions would still seem to suggest that the Sun goes around the Earth.

The nonexistence of the spheres wasn't even a particularly strong argument in favor of the Copernican system. Had the spheres, in fact, been physical, a Sun-centered system would indeed be impos-sible. But all their absence did was make a Sun-centered system *not* impossible.

The math, though, was compelling: Copernicus's math was clearly superior to Ptolemy's in terms of capturing the heavens on paper. It wasn't perfect; Copernicus's math still couldn't account for some subtleties in the motions. But when one of Johannes Kepler's professors at Tübingen University explained the Ptolemaic system side by side with the Copernican system, Kepler didn't need much convincing. "How," he wondered, "would the phenomena occurring in the heavens appear to an observer stationed on the moon?" The answer to this question Kepler hoped to make the topic of his disputation, the public defense of a dissertation in the German academic system.

Lutherans, however, were less accommodating of a Sun-centered cosmology than their more traditionally Christian counterparts. Martin Luther and his intellectual advisor Philip Melanchthon, citing biblical precedent, prohibited the teaching of the Copernican theory. In the Old Testament book of Joshua, when a battle between the Israelites and the Amorites turns in Israel's favor, Joshua asks God to make the Sun stand still so that his army can complete its rout before dark. God accedes to the request: "The Sun stopped in the middle of the sky and did not hasten to go down for about a whole day." Case closed: A Sun that God can make stand still during a battle is a Sun that, on all other occasions, must be in motion.

By the time Kepler was attending university, Luther and Melanchthon were long gone (Luther died in 1546, Melanchthon in 1560), but the Lutheran tradition of avoiding speculation was sufficient to convince the professor who would have been in charge of Kepler's disputation to refuse to allow it. Kepler put the manuscript in the proverbial drawer.

His allegiance to the Copernican interpretation of the heavenly motions, however, did not waver. In 1600 the great Tycho Brahe himself invited Kepler to join him at the observatory in Prague. By now Tycho was the Imperial Mathematician, appointed by the Holy Roman Emperor, Rudolph II. Tycho had at his disposal his vast collection of meticulous, multi-decade observations from his time at Uraniborg — and an allegiance to propriety that, during a royal feast, anchored him to a seat at the banquet table long after his bladder could contain his loyalty. He died ten days later, in October 1601, of a urinary infection.

To Johannes Kepler passed the title of Imperial Mathematician as well as the responsibility for finishing the Rudolphine Tables, a catalogue named for his royal benefactor that would rely on those meticulous, multi-decade observations of Tycho's to predict the positions of stars and planets. In 1604, Kepler had his own encounter with an anti-Aristotelian phenomenon when another nova appeared in the night sky, and then, for more than three weeks, in the day sky, too.

This evidence was no more conclusively Copernican than Tycho's 1572 nova and 1577 comet. But just as Tycho's interpretation of those phenomena differed from past interpretations of similar phenomena by virtue of the interpreter's superior knowledge of the empirical data, so did Kepler's interpretation of the nova of 1605. In Kepler's case, though, the interpretation depended not as much on the observer's observational acuity as on his math.

To accommodate vagaries in the circular orbits, Copernicus had placed the center of the system near, but not in, the Sun. Kepler realized that this off-centeredness actually made sense if you assumed that the shape of the orbits wasn't circular but elliptical. In an ellipse, the Sun could occupy one focus. The Sun would no longer be in the center of the system in the sense of always being the same distance from each orbiting object. But it would be central in the sense of being what every planet was orbiting.

In 1609 Kepler published his results in *Astronomia nova,* or *The New Astronomy* — an ambitious title that the book actually earned. For the first time in the history of astronomy, the math matched the heavens, and the heavens matched the math. No fudging necessary: no circles, or circles within circles — no saving the phenomena. The book demarcated the history of astronomy into a definitive *before* and *after:* what astronomy has meant since the beginning (or beginninglessness) of time versus what astronomy would mean forevermore.

Even so, the evidence was still merely compelling, not conclusive — a point that Kepler's friend Wacker von Wackenfels, a diplomat and royal advisor dabbling in astronomy, made in a series of conversations with him that same year. The focus of their discussion was the light and dark patches on the surface of the Moon. To Kepler's assertion that they were mountains, valleys, and plains, the emperor Rudolph himself had responded, echoing a common argument among the aetherists, that the irregularities were not on the surface of the Moon but were instead reflections of the Earth's own irregularities on the Moon's perfect surface.

Kepler's response was to reach back into the proverbial drawer and resurrect his Tübingen disputation of 1593. His dismissal of the aether, of the centrality of Earth, of spheres physical or metaphorical, would still be speculative, technically. But now his logic had a tactical advantage: the mathematics of elliptical, non-uniform orbits. What *Astronomia nova* did for astronomy, he hoped his tale would do for mythology.

Kepler borrowed the traditional narrative trajectory of sending his hero on a journey into the celestial realm and then having him turn his gaze back toward Earth; Kepler even named his story *Somnium,* echoing Cicero's tale of the two Scipios. Kepler, however, didn't need to dispatch his hero to the outermost edge of the universe, because his purpose wasn't to impress upon his readers their cosmic insignificance or the majesty of the heavens. Instead, Kepler sent his hero only as far as the Moon. He took his character to the border between the sublunar and the celestial, to the previously impenetrable membrane separating *down here* from *up there* — and then, as if his hero were arriving at a new shore, Kepler had him cross it.

Arriving at a new shore was what Kepler wanted the readers of *Somnium* to do. Arriving at new shores, if only in their imaginations, was what Kepler knew his prospective readers had already been doing, because they were all living in an era in which they awaited word from new shores — from the European navigators who were regularly "discovering" new lands, then returning with reports that put their own shores into a fresh perspective.

Since the 1420s explorers had been stepping onto new shores to the south, first beyond the western bulge in Africa, then around the southernmost cape and up the east coast of the continent. Since the 1490s explorers had been stepping onto new shores to the west, the islands and great landmasses that protruded from the ocean roughly midway between Europe and China. All these explorers were routinely returning with accounts of peoples and creatures and vegetation — and even two landmasses great enough to be continents — that were foreign yet familiar.

Kepler's dissertation, which of course had been an academic

work, would form the core of *Somnium,* but to that core he added a narrative framework, bookends to buttress its credentials as a work of fiction. In 1600, Kepler had to flee Graz because he wouldn't convert to Catholicism. In 1609, in Lutheran Germany, you still couldn't be too careful, and Kepler was quintuply so, at least. In his tale: A narrator, who might or might not be the author, tells the reader a dream, in which a narrator is reading a book, in which the main character, Duracotus, recounts a tale, in which Duracotus's mother, a witch, summons a daemon (spirit) to describe a certain phenomenon, in which daemons regularly transport people from Volva (read: Earth) to Levania (Moon).

The rest of the book, except for an *And-then-I-woke-up* final paragraph, consisted of details of life on Levania. Kepler populated his surrogate Moon with his own fanciful analogs to the Age of Discovery. Levania, he wrote, is home to two peoples (if that's the right word*): the Subvolvans, who inhabit the hemisphere that always faces Volva (Earth), and the Privolvans, who inhabit the hemisphere that always faces away from Volva. "The Subvolvan hemisphere compares very favorably with our cantons, towns, and gardens," Kepler's daemon explains, "while the Privolvan resembles our fields, forests, and deserts."

For Kepler, though, the inclusion of these details functioned in much the same way as the tale-within-a-book-within-a-dream structure of the story. It was a diversion—a misdirection that al-

---

* Probably not: The Privolvans "traverse the whole of their world in hordes, following the receding waters either on legs that are longer than those of our camels, on wings, or in boats."

lowed him to ease his readers into the tale's true purpose: not only to step onto another shore, look around, and see a world foreign yet familiar, but to step onto another shore and *turn* around. To consider the shore they'd left behind. From that perspective, wouldn't the old shore seem familiar yet foreign?

That familiarity was crucial for Kepler's program. The similarity between the Americas *over there* and Europe *over here* provided an implicit precedent for Kepler's true purpose: to evoke a similarity between *up there* and *down here.* If you were standing on the Moon, the Earth would be just one more planet. Which is what it was in the Copernican system. Which is what it was in Kepler's aborted dissertation. *Somnium* was a fantasy, but it was also a lesson in astronomy — the astronomy of the *astronomia nova.*

The fixed stars are so far away that they appear the same from Levania, but not so the nearer celestial objects. "One observes very many movements and numbers of planets different from those which we see from Earth, so that all of their astronomy has another meaning." The daemon goes on to provide lengthy explanations of the durations of day and night, the appearances of eclipses, and the positions of planets. Stand on the Moon, and you might think that you were at the center of the universe, and that the Earth is a planet in motion.

The historical record was not without philosophers speculating on the possibility of landmasses in the Atlantic Sea between Europe and China. The ancient Roman philosopher Strabo had written, in his *Geographica,* that "it may be that in this same temperate zone" — the zone near the equator that is conducive to human habitation — "there are actually two inhabited worlds, or even more."

Likewise, some philosophers had speculated about the existence of new worlds — *worlds* in the planetary sense. A disc of light dominates our daylight hours; dots of light dominate our nighttime hours. Might the two sources of light be the same type of phenomenon? Might the dots be suns? If so, might those dots of light and our disc of light possess the same properties — for instance, the attendance of orbiting planets?

Kepler himself found inspiration in the ancient Greek essayist Plutarch's *De facie quae in orbe lunae apparet,* or *Concerning the Face Which Appears in the Orb of the Moon,* a work that presupposed that the Moon was not a member of the exotic species of aethereal objects in the celestial realm but an Earth-like object. *De facie* was, Kepler thought, "the most valuable discussion of the earth's satellite to come down to us from antiquity." But Plutarch's Moon was still as firmly mythological as the heavenly spheres: It was the place souls go in the afterlife.

Kepler's *Somnium,* however, benefited from an advantage that no previous work of speculative literature could claim: the math in his own *Astronomia nova.* Dante had invoked a realistic interpretation of nature when he pressed his characters so far *down here* that they eventually found themselves going *up there.* Kepler, though, wasn't appending logic to a myth. For Kepler, the myth was literally an afterthought — as in eighteen years after. His tale was "speculative" only in the sense that daemons don't really ferry humans to the Moon, just as Virgil doesn't really guide Italian poets through Hell. Instead, Kepler was appending a myth to logic. The nonspeculative part — the astronomical part — was the point of the exercise. The math gave Kepler the freedom to consider the Sun as the center of the universe — or if not the exact center, then at least the central figure.

In the end, though, Kepler put *Somnium* in the proverbial drawer as well. The heliocentric model had already been a distinct possibility, thanks to Copernicus's math, and in Kepler's math it became a near certainty, but as a practical matter, what was *really* up there would have to remain what it always had been: a continent unexplored, an unreachable shore.

In April 1610 Kepler heard the news: We'd arrived.

An Italian mathematician, Galileo Galilei, reported that he had observed the heavens through a new instrument that magnified the powers of human observation such that distant objects appeared to be near. Through this instrument he had observed many wondrous sights in the night sky, each more magnificent than the last, each with its own significance for our understanding of where we stand in relation to the rest of the universe. Any of these observations would have changed the course of philosophical thought — previously unseen stars, moons around Jupiter — but one in particular would have especially pleased Kepler, a discovery that answered the ancient question of whether *up there* was the same as *down here*: mountains on the Moon.

The ancient conversation, it turned out, was only just beginning.

GRAVITY IN MOTION
3

As a student at Trinity College in the early 1660s, Isaac Newton kept two notebooks. In one he detailed his offenses against God dating back to his childhood and continuing through his current time at Cambridge University. The latter infractions included taking baths and making pies on the Lord's Day, as well as having "uncleane thoughts words and actions and dreamese." In the other notebook, *Quæstiones quændam Philosophicæ*, he addressed "certain philosophical questions."

The two subjects — God and nature — were never far apart in Newton's mind. While nothing in nature could exist without God, Newton wanted to know to what extent the somethings in nature — matter — could influence one another — motion — without the immediate intercession of a divine power.

"Of Gravity & Levity," Newton wrote at the top of a fresh page in his notebook on nature. He was using the terms *gravity* and *levity* in the same manner as the ancient Greeks: the upward and downward tendencies, the effects that we observe in a stone's fall or a flame's rise. What concerned Newton was the cause of those effects — the matter causing other matter to move.

Levity was of less interest to Newton, maybe owing to an unthinking extension of our own human circumstances relative to the surface of the Earth: Why we remain on the surface is more inherently compelling than why we're not rising. "The matter causing gravity," he began, "must pass through all the pores of a body." Why pores? Because if the cause of downward motion acts on an object, it must act everywhere on the object equally. It must infiltrate the body of matter, cause its effect, and then move on.

But move on where? When it's done causing a stone to fall to Earth, this matter has to go somewhere. Perhaps it goes even farther down, and in that case, it must be collecting in the Earth, and in *that* case, it must be doing one of two things: continuing to fill vast underground cavities, cavities it would have been filling since the Creation; or swelling the Earth itself. Neither seemed plausible to Newton.

Not down, then. So: up.

But an answer of *up* presented another problem. After a material cause forces the same object downward, it can't then ascend in the same form in which it descended; if it did, it would force a body upward. Therefore the cause must change form, and with it, capacity to influence.

But change from what to what?

Newton hadn't a clue. Yet this material cause must be there. So whatever it was, in either its *going-up* or *going-down* guise, it must be—

"Gummy," he guessed, in a 1675 letter to the theologian and philosopher Henry Oldenburg. "Something very thinly & subtily diffused." Also: "unctuous." And: of a "tenacious & Springy nature."

In other words—and words are part of what would continue to fail him throughout his decades-long struggle to articulate the motions of matter—Newton still didn't have a clue.

The story goes that he saw an apple fall. The Black Plague having chased him from Cambridge in 1665, Newton was strolling on his mother's farm in Woolsthorpe, and he saw an apple drop to the ground, and he asked himself: Could the same cause of the motions of matter—whatever it is—apply to an apple as to the Moon?

Newton didn't need a lesson from fruit.* Knowing what he already knew, he couldn't *not* have assumed that the matter *up there* was the same as the matter *down here,* and therefore *not* have wondered whether the same principles of motion might apply in both realms. Philoponus had helped secure in the surviving canon of ancient writings the possibility that matter was the same in the heavens as on Earth, as part of his speculations on a Creation that brought into existence all matter all at once. Copernicus had transformed

---

* Which doesn't mean he didn't get one. He just didn't need it.

through mathematics the possibility that Earth was a planet into a probability. And then Galileo had advanced the proposition from probability into near certainty with his observations through a telescope.

Over the course of a fifty-two-page pamphlet—the one that reached Kepler in the spring of 1610—Galileo had revealed more new information about the celestial realm than in all of human history. Having overcome his own confusion and disbelief, he realized he would need to overcome his readers' confusion and disbelief if he were to ease them into the new universe—one that he had discovered, and one that he, for the moment, was the only human to inhabit.

Galileo began with the instrument that had made the information available. After hearing rumors and descriptions of a new type of spyglass through which distant objects appeared nearer than they actually were, Galileo wrote, he figured out how to make such an instrument himself: He outfitted a lead tube with a lens at either end, one concave, one convex. He led readers through an explanation of how they, too, could make such an instrument. Then he recounted sights that were "hardly believable," even "inconceivable."

He saw mountains on the Moon. He'd watched as the peaks of the mountains blossomed out of the darkness into a new dawn, followed by the rest of the mountains as the Sun continued, from the lunar perspective, to "rise." He had measured the changing lengths of the shadows. He had drawn illustrations that left little to the imagination.

He saw stars. To the Pleiades cluster of six stars, Galileo added forty; to the constellation of Orion he added five hundred, including eighty in the belt and sword alone. He resolved the foggy features of previously nebulous regions of the night sky into multitudes of individual stars; he concluded, for instance, that the Praesepe nebula is not one fuzzy object but a collection of forty distinct stars. He did the same with the Milky Way itself — the spill of white that arcs across the sky. The Milky Way is, he noted, "nothing else than a congeries of innumerable stars distributed in clusters. To whatever region of it you direct your spyglass, an immense number of stars immediately offer themselves to view, of which very many appear rather large and very conspicuous but the multitude of small ones is truly unfathomable."

He saw four moons around Jupiter. This particular discovery was in fact the reason Galileo had rushed the pamphlet into print; he had first observed them in January 1610, and *Sidereus Nuncius*, or *Starry Messenger*, appeared that March 13. The moons of Jupiter proved that the Earth isn't unique as a center of rotation. Once again, the evidence wasn't conclusive in favor of the Copernican system, but as was the case with the impossibility of physical spheres, the presence of at least two centers of rotation (Earth and Jupiter) removed an argument in favor of the alternative. Late that same year, Galileo made the series of observations that even the Wackenfelses of the world couldn't explain away: the phases of Venus as the planet orbits the Sun.

As the technology of telescopes improved, other astronomers discovered more matter *up there* — moons orbiting and rings encircling Saturn, for instance. But they also began to detect a telling pattern among the *motions* in a celestial realm they increasingly could imagine as encompassing *down here* as well. The planets, including Earth, not only orbit the Sun, but they do so in a uniform style. They lie in a plane, more or less, much like the moons of Jupiter or the rings of Saturn relative to their "host" planets. To the French philosopher René Descartes, writing a couple of decades after Galileo's first observations through the telescope, this uniformity or nesting effect suggested either that each body was swirling within a so-far-invisible substance, as if caught in a funnel or vortex, or that all the bodies were swirling within that substance, as if caught in funnels or vortices.

The assumption behind this principle — that an invisible yet physical presence is the cause of motions — was the one that Newton had applied in his notebook of philosophical questions. He was looking for some sort of immediate connection between object and object, between mover and moved. Only when he turned his attention to the motions of matter *up there* did he begin to make some progress, because only then did he set aside the question *What cause creates the motions down here?* Instead he began to pursue the ancient question *What math matches the motions up there?* In particular, Newton wanted to know: Does Kepler's math match those motions?

After years of examining the observations of Mars that Tycho had left him, Kepler had noticed a mathematical relationship: The position of the planet over the course of an orbit seemed to correlate with its velocity. As Mars nears the Sun, its velocity increases, but as it recedes from the Sun, its velocity decreases. On closer examination, Kepler found that for each unit of distance away from (or toward) the Sun that Mars travels, its velocity drops off (or increases) by a certain rate. At double the distance, the velocity is one-fourth;

at three times the distance, it's one-ninth; at four times the distance, it's one-sixteenth; and so on. In mathematical terms, the distance and the velocity have an inverse-square relationship.

The ancients knew that the planets do not go about their business at a uniform rate; sometimes a planet seems to be speeding up, sometimes slowing down—one of the observations that encouraged them to resort to other geometrical means in order to save the phenomena. Kepler himself had suggested the shape of Mars's orbit, but it was only one of several geometrical options available to him. "Kepler," Newton wrote, "knew the orbit to be not circular but oval" —an elongated circle—"& guest it to be elliptical"—a particular kind of oval. Kepler had also stipulated—guessing again—that if Mars travels in an elliptical orbit, then all the planets travel in elliptical orbits. Despite these uncertainties, Newton was willing to give Kepler the benefit of the doubt. Let's say Kepler is right, Newton reasoned. Let's assume that the shape of planetary orbits actually *is* elliptical. Now what?

Now this: the math.

The universe is "written in the language of mathematics," Galileo argued in his 1623 book *Il Saggiatore*, or *The Assayer*, challenging the largely observational-logical philosophical method he'd inherited. Thanks to the telescope, mathematicians now had access to measures of motions that were not approximate but accurate. In which case, Galileo argued, the goal of mathematicians no longer should be to save the phenomena—to find the math that most closely matches the motions. Instead their goal should be to find the math that *does* match the motions, because one mathematical description must.

Newton agreed, as did his peers who were trying to write down the rules of planetary orbits in the language of geometry. But unlike Galileo, and unlike his peers, Newton could speak in tongues—a mathematical language he invented out of necessity.

An orbit is not the result of one point of contact between a cause and matter followed by another point of contact between a cause and matter followed by another and another and another, in some sort of harmonic convergence of collisions. It is not a series of effects proceeding from a series of causes. It is, instead, a *continuous* effect—single and smooth. That continuity was sure to frustrate a purely geometrical approach. Mathematicians couldn't just measure the angle to the Sun over here, at this point in the orbit, and then again over there, at that point in the orbit, and compare the two, as if the angles were discrete effects of discrete causes. The mathematicians needed a conceptual tool that would allow them to take the measure of matter in one continuous motion. And Newton had that tool, because he'd created it.

Calculus allowed Newton to divide an orbit—or any curve—into as many small increments as the universe allowed: an infinite number of infinitesimal triangles.* After he did so, he found that an elliptical orbit didn't just result from an inverse-square model. It demanded one. The elliptical effect could arise from no mathematical cause other than an inverse-square model.

Problem solved. And the problem having been solved, Newton set it aside.

———————

* Calculus also would have allowed Philoponus's critics to counteract his own criticism of Aristotle's idea of infinity.

In the summer of 1684, Newton — by now a Cambridge professor and philosopher, as well as a mathematician of long standing and great repute — received a visit from Edmond Halley, an equally renowned astronomer and mathematician. Halley mentioned a question that had recently been vexing several of his acquaintances in London: What geometrical shape describes a planet's orbit around the Sun? "Elliptical," Newton answered,* to which Halley sputtered, "But, but— Surely you jest! Surely you haven't already—? Good God, man!"† To which Newton said, yes, he'd already done the math, and yes, he had the proof here somewhere, among his papers, but he couldn't place his hands on it at the moment. Keep looking, Halley urged. Newton did, and he found it.

The application of calculus had given Newton an invaluable, insurmountable advantage over those of his peers who were also working on the problem of orbital shapes. But so had Newton's intuition. While Halley and his friends were asking what orbital path an inverse-square relationship would dictate, Newton had wondered what would happen if Kepler was correct: What relationship would dictate an elliptical orbit?

But working through the math this second time, at Halley's request, Newton had a further insight. He needed to reach one hundred years into the past in order to retrieve the relevant prompt, and then he had to refine it for his own purposes. But as with many deeply profound insights, it was also deeply simple, once you saw it.

———————

* Not a direct quote.

† Ditto.

Never mind for the moment the cause of motions — the question that had vexed Newton at Cambridge and for many years afterward. Never mind the mechanics of motions — the orbital shape he'd determined to be elliptical. Never mind the math of motions — the calculus that told him an orbital shape could *only* be elliptical.

Instead, mind *motions*.

Consider the concept as a category all its own. Strip the category of the distractions that come naturally to anyone studying motions: what makes them happen or what shape they take or what laws they follow. Ask instead — as Galileo eventually did, and as Newton, following Galileo's example, now asked — a far more fundamental question: What does *to be in motion* even mean?

Galileo came of intellectual age in the *Is Earth a planet?* era of philosophical debate. He himself opened the *Earth is a planet!* era, beginning with his 1609 and 1610 discoveries of mountains on the Moon and of Jupiter's moons. One of Galileo's tasks, as he understood his historical moment, was to ease the transition from one era to the next by explaining not just the motions of matter but the motions of one particular piece of matter: our planet.

The idea that the Sun, not the Earth, might occupy the center of the universe, and that the Earth completed one rotation on its axis every day and one orbit of the Sun every year, wasn't new. It had appeared occasionally in the writings of the ancients, but it had always met with two objections. First, if you toss a rock straight up into the air on a moving Earth, it would have to fall behind you or ahead of you or off to your side or on any other patch of ground, depending on which direction the Earth was turning. The one place it wouldn't land, assuming you hadn't budged, was your hand. And the distance it landed from your hand would correspond to how fast the world was turning, wouldn't it? The second problem was that on a moving Earth, objects — including you — would be flying off the surface, wouldn't they?

No and no — or so Galileo suspected. He began building his argument by starting where seemingly everybody in history had started: simple, vertical, straight-down motion. Aristotle had assumed, just as his predecessors had, that the rate of an object's fall would depend on its weight — the heavier, the faster; the lighter, the slower. Galileo, though, wondered why either assertion should be true.

Take the heavy object, Galileo argued, and divide it in half. Then tether the two halves together. Won't they fall at the same rate as the heavy object? Yet each is half the weight of the total, so by Aristotle's reasoning each should be falling at a slower rate. Galileo's logic told him this possibility doesn't make sense: Under ideal conditions — without any air resistance, or, better, in a vacuum — the two halves and the whole would fall at equal rates.

To study this assertion, Galileo would need to overcome a limitation in the human sense of sight: Objects fall faster than the eye can see — or at least faster than the eye can observe closely enough to make accurate measurements using the instruments available to Galileo. So he had to slow down the actions to an observable form. His solution was to take the vertical, straight-down path of a falling object and angle it a bit. He constructed finely shaved wooden chutes down which he could roll balls. Galileo could vary the weights of the balls, and he could vary the angle of the chutes. No matter what choices he made, though, the key result was always the same: Balls of different weights reached the bottom of the chutes at the same time.

The experiment was successful. Galileo had asked nature a question, and nature had responded with an answer. But the experiment provided a bonus that Galileo hadn't anticipated: two more insights about falling objects.

A ball always *picked up speed* during its descent. At whatever angle a ball began to roll, and with whatever speed, its velocity always increased along the way. It didn't start rolling at, say, one inch per second and stay at one inch per second. It sped up.

But not only did it speed up, it sped up *at an unvarying rate*. At whatever angle a ball began to roll, and with whatever weight, its velocity always increased according to a mathematical relationship. After one second of rolling, a ball would have covered one unit of distance — one inch, maybe, or one foot, or one meter. After two seconds, though, the ball would have covered four times that unit of distance: four inches, four feet, four meters. After three seconds, it would have covered nine times the distance. After four seconds, sixteen times; after five seconds, twenty-five times; and so on.

And if this relationship holds when the incline is shallow, and still holds when the incline is not so shallow, and continues to hold when it's at an angle you might even call steep, then this relationship would still hold when the incline is as steep as it can be: vertical. Even though the ball would be moving faster than the eye can measure, Galileo could now be confident that it was still gaining speed at the same rate: The distance would be proportional to the time squared.

Having extrapolated from an inclined roll to a vertical fall, Galileo then reversed his reasoning. Rather than raising the incline and then raising it some more until it reached pure verticality, he could lower it and then lower it some more until the incline was as low as it could go: horizontal.

At this point Galileo enjoyed yet another insight, this time not about falling objects but about rolling objects. When a ball had finished its journey down a chute and reached the bottom, it would continue to roll. Now, though, it would no longer be gaining speed. Free of whatever had been causing it to accelerate during its descent, it would be rolling at a *constant* rate — or it would if the ball was perfectly round and the plane upon which it was rolling was perfectly smooth: ideal states that don't exist in nature, just like the resistance-free environment that Galileo imagined for falling objects of different weights. Even so, the principle behind this insight of rolling at a constant rate gave Galileo license to recognize one of the unthinking assumptions of the ancients, and to rethink it.

The ancients had wondered what kept matter in motion once it started moving. According to Aristotle, if you let go of a stone, it will begin its descent to the center of the universe as soon as it loses contact with your hand. And it does drop straight down, if all you do is let go. But what if you propel it — toss it, throw it? It will still apparently lose touch — literal touch — with anything outside itself, yet it won't fall straight to the center of the universe. Its motion will be non-straight — *violent,* as the ancients called it. Therefore, despite appearances, it must still be in touch with something that is moving it. But what?

The air, Aristotle suggested. What else is in the projectile's vicinity? The air must remain in contact with the projectile, keeping it aloft, because the air is the only matter coming into contact with the projectile. But how does the air keep the projectile aloft? Aristotle proposed that as the projectile moves forward, it pushes aside the air directly in front of it, and this air in turn passes to the side and rear of the projectile. In this way, air remains in constant contact with the projectile, compelling it to continue its forward motion.

Philoponus, characteristically, spared no contempt. "These things," he wrote, "are entirely unbelievable and are more like fictions." Over the centuries Aristotle's the-air-does-it arguments acquired a reputation as among his most questionable. Aristotle himself, if you read between the lines, regarded his hypothesis with something approximating a wince and a sigh: *If posit I must . . .*

But if not air, then what? Nothing, Philoponus answered: "It will not need something pushing from outside." Instead, the source will be *internal.* "It is necessary," Philoponus wrote, "for some incorporeal power to be given by the thrower to the thrown thing." The projector — the bow for the arrow, the hand for the stone — must endow the projectile with a momentum of some sort. Projector and projectile will lose physical contact with each other, but they'll remain in virtual contact through a transfer of power.

When Galileo began his own investigations into the motions of matter, Philoponus's internal transfer of power was a frequent source of inspiration. The sight of a ball rolling on a smooth plane, however, inspired Galileo in another direction. Looking in his mind's eye at ideally round balls rolling on an ideally smooth surface, he could see that once in motion they would keep on rolling. Just as Philoponus had removed Aristotle's external influence and replaced it with an internal influence, so Galileo now removed Philoponus's internal influence and replaced it with . . . nothing. Rather than ask what keeps matter in motion, Galileo inverted the question. He asked what made matter in motion stop.

What does it mean to be in motion? We, observing a ball, see it rolling. But what does the ball "see"? From the point of view of a ball rolling on a frictionless plane at a constant rate, Galileo reasoned, there would be no difference between being in motion and not being in motion.

In his 1632 *Dialogue Concerning the Two Chief World Systems* — the two world systems being the Aristotelian and Copernican — Galileo used this insight to explore how a stone thrown straight up doesn't land far away. Imagine you're on a ship, standing inside a windowless compartment. If the ship is moving at a constant rate, everything in that compartment will also be moving at the same constant rate. You, inside that cabin, would have no way of knowing whether you were moving or not. You yourself would be feeling no push or pull, and neither would anything in your environment. A fish in a bowl would not be slapping against the glass. The water in the bowl would not be staining the table. If you were to drop a stone, it would fall straight to the floor.

Now remove the walls of the cabin. Look — you're moving! Either that or — an equally valid interpretation, according to Galileo — you're standing still and the shore is moving. Yet the fish is content and the water in the bowl is calm and the stone still falls straight to the floor because you and the bowl and the fish and the stone are moving as one — a single unit.

Take that principle and apply it to the Earth. If you're standing on the surface of the Earth and dropping stones "straight" down or throwing them "straight" up, they'll fall at your feet, because you and the stone and the Earth are moving at the same speed, allowing you and the Earth to move as a single unit.

The principles of motion that Galileo had reconceived involved the matter *down here*. Newton, trying to answer Halley's question, saw no reason the same principles shouldn't apply to matter *up there*. The matter down here and up there was very (very, very) likely the same; maybe the principles behind the motions were the same. In that case, Newton asked, why should a planet in motion not keep going straight ahead. Why should it stop?

Well, not *stop*. But bend. Curve. Depart from its straight path and adopt an elliptical one.

Newton was going to need a bigger book.

❖

Newton's re-creation of his own calculations had far exceeded Halley's challenge. Like Galileo asking one question about falling objects and receiving, for his labors, a host of insights, Newton had asked Halley's question about planetary orbits and was now unearthing—maybe not the right word, since he wasn't tilling the ground down here, even metaphorically; but then again, maybe he was, if *down here* and *up there* possessed the same physical properties—a bounty of his own.

The work, Newton decided, would consist of two volumes, and

he considered calling them, collectively, *De motu corporum,* or *On the Motion of Bodies.* No longer was he worrying exclusively about identifying the mysterious something that causes the effect we call motion. He was wondering instead about motion itself. In writing *De motu,* Newton elevated the most important of Galileo's key, if unsought, insights to a formal principle:

Matter at rest will remain at rest unless moved. Matter in motion will remain in motion unless stopped.

In the Copernican/Galilean interpretation of the universe, Earth is a planet. We can therefore think of it as we do other planets: as matter in motion. According to this principle, though, Earth is not just matter in motion. It's matter in motion that will remain in motion until an obstacle stops it. Which is to say, Earth is a projectile.

The idea of planets (though not Earth) being projectiles was not original. Philoponus had argued that if terrestrial objects move via a transfer of power, and if celestial objects are no different from terrestrial objects, then the heavens move via a transfer of power. And this transfer of power, Philoponus added, occurred at the moment of creation, and it is what keeps the heavens in motion to this day.

Philoponus's argument influenced the fourteenth-century French priest and philosopher Jean Buridan. "Since the Bible does not state that appropriate [angelic] intelligences move the celestial bodies," Buridan wrote, in his *Questions on the Eight Books of Aristotle's Physics,* "it could be said that it does not appear necessary to posit intelligences of this kind. For it could [equally well] be answered that God, when He created the world, moved each of the celestial orbs as He pleased." Throw a stone and it carries forward with the *impetus*—the name Buridan gave it—that you've imparted to it. Throw a planet and it carries forward with the impetus that God imparted to it.

Newton, however, no longer had recourse to the concept of impetus. Instead he had to incorporate matter in motion staying in motion unless stopped. The philosophers who had regarded a heliocentric model of the cosmos with skepticism had been right all along, though not for the reason they thought. Objects *should* be leaving the surface of a spinning Earth. You are a projectile, and you *should* be flying off.

And you would be, if the only motion were the straight-ahead, forward motion—the one that Galileo had described and Newton had elevated to a formal principle. But you don't fly off. Therefore another motion must be at play, and Newton realized which one: the downward motion that Galileo had measured with his inclined planes and falling balls. Put those two motions together—the forward, or *inertial*, and the downward, or *centripetal* (a coinage of Newton's meaning "falling toward the center")—and you have the appearance of a single unit in a single motion. And that situation will persist—objects on the surface of a spinning planet will remain on the surface—as long as the downward motion is stronger than the forward. If the forward motion became strong enough—if the Earth's rotation were fast enough—then, yes, you would fly off.

Newton, however, wasn't exploring only the motions of the

Earth. Unlike Galileo, he wasn't trying to explain only how the Earth could be a planet. He wanted to explain the motions of planets, of which Earth was one.

For that reason he wondered if he could consider the Moon and Earth to be the equivalent of a single unit in motion. The Moon on its own, like you on the surface of the Earth, should be moving in a straight line, but it's not. Therefore, like you on the surface of the Earth, it might be experiencing another motion — the downward one. Would the same combination of two motions explain the orbit of the Moon? Does the Moon's straight-line motion curve — that is, does it turn into an orbit — because it is falling toward the Earth?

Galileo had assumed that the rate of downward acceleration he'd measured in his experiment would remain the same over time and space. Newton, though, figured that it would be subject to the same inverse-square relationship that Kepler had discovered relative to distance and velocity. Newton took Galileo's measurement of acceleration at the surface of the Earth, subjected it to the inverse-square law at the distance of the Moon, and factored in the Moon's velocity.

The math matched.

Newton then extended this principle. If it applied to the Moon's relationship to Earth, then it should apply to the moons of Jupiter and Saturn relative to their "host" planets. And if it applied there, then it should apply to the planets' relationship to the Sun. Newton was ready to make his pronouncement, and he did so in the ever-burgeoning volume *De motu*: "Therefore the major planets revolve in ellipses having a focus in the center of the sun."

But then came a potentially crippling realization: He had been treating the planets and the Sun as points in space. They aren't points. They're bodies full of matter — the matter that would have a downward influence on other bodies of matter. The Moon is falling toward Earth. But Earth is falling toward the Moon . . . even as it's falling toward the Sun . . . and the Sun is falling toward Earth . . . and Jupiter is falling toward Earth and the Sun and the Moon, and they are all falling toward Jupiter . . . and on and on and on, all the heavenly bodies traveling in their own straight lines while always falling toward one another.

Newton was going to need an even bigger book.

He set aside *De motu* and began work on a new volume. It would contain pretty much everything he'd written for *De motu* but with the further complication of what matter falling toward matter would do to the perfect regularity of orbits: make them imperfect (albeit in perfectly predictable ways). All that tugging and pulling will take a toll on each object, but the system as a whole will remain stable.

"The planets neither revolve exactly in ellipses nor revolve twice in the same orbit," he wrote. "Each time a planet revolves it traces a fresh orbit, as happens also with the motion of the moon, and each orbit is dependent upon the combined motions of all the planets, not to mention their actions upon each other."

Newton also realized that his logical progression was incomplete. He had begun with Galileo's ideas about rolling balls *down here* and extended the lessons to rolling balls *up there.* But if the principles of motion that held down here also held up there, then the principles that held up there had to hold down here. Meaning: All the matter down here interacts with all the other matter down here as well.

And because the divide between up there and down here is, if not arbitrary, at least artificial, then the matter down here interacts with the matter up there.

Everything in the universe is in a constant state of falling toward everything else, no matter the distance. Or, if you prefer, rising toward everything else — a distinction of psychological convenience, perhaps, but of no physical significance.

Newton would, of course, have to test such a preposterous yet logical series of propositions. Appropriately, he devised two such tests, one down here and one up there.

Down here, he applied his math to the tides. Newton realized that for his purposes he could treat Earth as, more or less, a globule of water. Like everything on a turning planet, the water has two motions — straight and falling. Like everything on a planet, the tides are also under the influence of everything else in the universe, an effect that is negligible for most bodies on Earth but not for something as vast and malleable as the oceans. Newton consulted voluminous historical collections of tidal data, combined them with the positions of the Moon and the Sun at those precise times, and applied the formulas.

The math matched. The Moon's influence on the tides turned out to be slightly more than double the Sun's, exactly what Newton predicted.

Up there, Newton put comets to the test. Tycho and Kepler had decisively demonstrated that comets emerge from a region beyond the Moon. But Halley noticed a pattern that suggested comets might return. In that case, did they orbit the Sun, and if so, did they follow elliptical paths? Newton assembled the same kind of voluminous observational data of comets as he had with tides, traced their paths, applied the calculations.

Again, the math matched.

Halley himself bankrolled the *Philosophiae Naturalis Principia Mathematica,* or *Mathematical Principles of Natural Philosophy,* which was published on July 5, 1687. He also reviewed the book, anonymously, in *Philosophical Transactions,* by far the most influential publication representing the new breed of philosophical investigation that was just emerging. The review was a rave:

> This incomparable Author having at length been prevailed upon to appear in publick, has in this Treatise given a most notable instance of the extent of the powers of the Mind; and has at once shewn what are the Principles of Natural Philosophy, and so far derived from them their consequences, that he seems to have exhausted his Argument, and left little to be done by those that shall succeed him. His great skill in the old and new Geometry, helped by his own improvements of the latter, (I mean his method of infinite Series) has enabled him to master those Problems, which for their difficulty would have still

lain unresolved, had one less qualified than himself attempted them.

Newton was no closer to identifying whatever was counteracting the straightforward, inertial motions throughout the universe — the something that *up there* foreordained orbits and *down here* kept people from flying off the planet — than when he first addressed "certain philosophical questions." The noun he had invented to describe the downward effect — *centripetal* — covered only the falling part of the overall motion. Newton needed a noun for the combination of the two effects, the inertial and the centripetal — the overall effect that we perceive as elliptical orbits and falling dirt. His choice was *gravitation*.

The word wasn't his invention. Variations had been around for thousands of years. *Gravity* was still a synonym for weight, as it had been for the ancients. *Gravity* was still what it had also become in recent decades, as more and more investigators of nature had begun to wonder how the world works: a word more or less meaning "the act of falling" or "how fast an object falls," as in "The gravity of the object was swift." Newton's *gravitation*, however, was new. It was not so much a description of an object's behavior as a *thing* worthy of consideration unto itself.

Newton knew he was asking a lot of readers, to believe in something they couldn't see, taste, touch, feel, or hear. And not just to believe in it, but to believe it to be the cause of the workings of the universe. And not just to believe it to be the cause of the workings of the universe, but to believe it to be the cause of the workings of the universe without knowing how it itself worked. Even God was easier to believe in. Even a God who cared whether you baked a pie on the Lord's Day was easier to believe in. At least that God was a matter of faith. But to believe in gravitation — as Newton also knew — required more than a leap of faith: a leap of fiction.

GRAVITY AS A FICTION

The criticism came from the greatest minds of the era, and it came at once.

The *Principia*, everyone agreed, was one of the most impressive intellectual feats in the history of civilization, a multivolume masterpiece more than living up to Halley's effusions, a nonstop tour de force of mathematical heights and philosophical depths. The consensus, also, was that it was missing something.

"I am not especially in agreement with a Principle that he supposes in this calculation and others," wrote Christiaan Huygens, the most influential natural philosopher on the Continent, in his 1690 tract *Discourse on the Cause of Gravity*, "namely, that all the small parts that we can imagine in two or more different bodies attract one another or tend to approach each other mutually. This I could not concede, because I believe I see clearly that the cause of such an attraction is not explicable either by any principle of Mechanics or by the laws of motion." The German philosopher Gottfried Wilhelm Leibniz — a former student of Huygens as well as a mathematician who had independently created a form of calculus — echoed this objection, and then over the years echoed his own echoes. "It is true," he wrote in 1710, "that modern philosophers for some time now have denied the immediate natural operation of one body upon another remote from it, and I confess that I am of their opinion."

Newton wasn't of their opinion. But he also wasn't *not* of their opinion. He agreed with his critics that what causes matter to deviate from a straight-ahead inertial motion can't be some mysterious interaction with other matter that involves no physical contact. "That Gravity should be innate, inherent and essential to Matter," Newton wrote to the young cleric Richard Bentley in a series of letters in the winter of 1692–93, "so that one body may act upon another at a distance thro' a Vacuum, without the Mediation of any thing else, by and through which their Action and Force may be conveyed from one to another, is to me so great an Absurdity that I believe no Man who has in philosophical Matters a competent Faculty of thinking can ever fall into it."

Yet fall into it was what his critics accused him of doing. They had their reasons. The ancients were able to say that whatever is *up there* stays up there because it's made of different matter from what's *down here* and therefore follows different rules. But Newton, first by accepting the assumption that the two realms consist of the same matter, then by assigning the same cause to their motions, had removed the luxury of not having to specify a cause. His critics assumed that he therefore must have specified one, and as best they could figure, it was attraction — whatever that means.

But Newton had not specified anything, other than an old noun with a new nuance. Even so, his critics were missing the point. Newton had not yet identified the mechanism that makes earth fall to Earth — the cause, which he called gravitation, of an effect, which he and everyone else called gravity. "The Cause of Gravity is what I do not pretend to know," he wrote at the end of that same letter to Bentley. What Newton did know and what his critics failed to appreciate — both upon publication of the *Principia* and for decades afterward — was that not only had he not assigned a cause to gravity, he didn't need one.

Newton had published the first edition of the *Principia* at the age of forty-four; when the second edition appeared, in 1713, he was seventy-one. He'd been busy. In the years since that initial publication, he'd written numerous religious tracts; presided over the Royal Society, an institution that represented the greatest minds of Britain; served as Warden, then Master, of the Royal Mint; and, in 1705, published *Opticks*, an investigation into the nature, properties, and behavior of light. But he'd also been his usual meticulous, prickly self, ignoring the entreaties of his benefactors to *hurry up, man* — a plaint that the editor of the second edition couldn't resist airing in public, albeit while phrasing it as a compliment. Toward the end of his preface, Roger Cotes, a professor of astronomy and experimental philosophy at Trinity College, praised the forbearance of the publisher

— the same Richard Bentley who twenty years earlier had corresponded with Newton. Bentley, Cotes wrote, had been trying "with persistent demands to persuade Newton (who is distinguished as much by modesty as by the highest learning) and finally — almost scolding him — prevailed upon Newton to allow him to get out this new edition, under his auspices and at his own expense."

Yet for all his frustrating behavior, Newton could, once a task caught his attention, lash himself with an intensity approaching mania. While investigating the nature of light for his *Opticks* he stared into a mirror reflecting the Sun for as long as he could (then shut himself in a dark room for three days, waiting for the spots to fade). While running the Royal Mint he donned disguises and frequented pubs in order to personally ferret out counterfeiters. Now he would apply the same single-mindedness to composing a scholium — a scholarly term for an elaboration on a text — that would conclude the second edition of the *Principia*. The idea of two objects interacting with each other across an expanse of space without anything tangible connecting them was no less an absurdity to him than it had been a quarter of a century earlier, and he was no closer to ridding his masterpiece of its central incoherence. He was also no closer to thinking that the flaw was of concern, let alone fatal to the whole enterprise — and he wanted his readers to understand why.

"Now something must be said about the method of this philosophy," Cotes wrote in his editor's preface, easing the reader into Newton's reasoning. Cotes began by providing a historical context, dividing the study of the motions of matter into "roughly three classes."

The first was the one that the ancients and their more recent emulators followed. They spoke of the "nature" of objects and their "natural" motions. They "endowed the individual species of things with specific occult qualities" — mysterious reasons that pretty much boiled down to *This matter moves in the manner it moves because that's how this matter moves.* By current standards, Cotes wrote, they wouldn't even be philosophers; they would be "inventors of what might be called philosophical jargon."

The second category was the one that Newton's generation had inherited and Newton himself had abandoned. The practitioners of this philosophy, Cotes said, "take the liberty of imagining" the existence of particles of "unknown shapes and sizes" that possess "uncertain positions and motions" determined by "occult fluids that permeate the pores of bodies very freely." These particles would, for instance, make gravity happen; they're the kind that young Newton assumed must be present. (Think "gummy.") But they are, in the end, the product of philosophers "drifting off into dreams." Even if those philosophers then "proceed most rigorously according to mechanical laws" — even if they then adhere to the mathematical equations that capture the motions of matter — they are "merely putting together a romance, elegant perhaps and charming, but nevertheless a romance."

The third class was the current one — the class for which Newton had become the foremost exemplar, the method for which gravity had become the greatest test. Its practitioners called it the New Philosophy and themselves New Philosophers, precisely because even as they continued ancient traditions of philosophy, their methods were more ambitious and more trustworthy than any precedents.

As recently as the twelfth century, or the fourteenth, or the sixteenth, scholars could think of themselves as studying a monolithic body of knowledge. Not adding to it, because there was nothing to add. It was complete, and it was available, and it went by the name of the works of Aristotle. Scholars regarded Aristotle's writings as a secular bible: The Word of God told you everything you needed to know about the spiritual world, and the word of Aristotle did the same for the natural world. But that body of knowledge proved to be not so much a monolith as a shibboleth. Never mind Aristotle's misdiagnoses of motions in the sublunar realm. In the heavens alone, his insistence on the presence of celestial spheres, the ubiquity of uniform circular motion, and the centrality of the Earth had revealed his (understandable) ignorance.

So if not Aristotle in ancient times, then who and when? The knowledge of nature must have already undergone corruption before Aristotle inherited it. Surely, in some remote and possibly irretrievable past, mankind must have solved the riddles of nature. Eden-like, that epoch had existed in a state of intellectual perfection. In which case, the body of knowledge *would* be monolithic. It just wouldn't be the current body. It wouldn't even be a Platonic body of knowledge — a set of facts out there in some ideal form that was approachable but ultimately unreachable. Instead, it would be a *lost* body of knowledge.

Because what was the alternative? What other interpretation of their historical situation could there be? That they were living at a moment when the world was ready not to rediscover but to *discover* the secrets of the universe? That *now* was special? That history had endowed *them* with the opportunity to see sights heretofore unseen, to think ideas heretofore unthought? Such an assertion would be presumptuous in the extreme, a sin of pride even, if only because, well, *Why us?*

Certainly the knowledge of the world and the universe had grown in ways the ancients could hardly have imagined: sights heretofore unseen. New worlds *up there,* visible through the telescope. New worlds *down here* — not just the continents and peoples on the other side of the oceans but the multitudes living in our spittle and our soil, visible through the microscope. Knowledge of human anatomy that arose not from guesswork but from gutwork, so to speak: by getting up-to-the-elbow bloody.

The English physician and philosopher William Gilbert had been the first to explain this hands-on method. "In the discovery of secret things, and in the investigation of hidden causes," he wrote in his *De Magnete,* or *On Magnetism,* in 1600, "stronger reasons are obtained from sure experiments and demonstrated arguments than from probable conjectures and the opinions of philosophical speculators."

Sure experiments; demonstrated arguments: evidence; logic.

The ancients had relied upon a combination of evidence and logic, but their evidence, such as the circular motion of heavenly bodies, had turned out to be unreliable and their logic pretty much what you might expect when starting from incorrect assumptions. The New Philosophers, however, saw little value in adopting *any* approach until they could satisfy themselves that they weren't heading toward a lethal crossroads of wrong information and weak rationales. And the only way to ensure that they wouldn't repeat the errors of the past, they felt, was to begin again. They would start the study of nature from first principles.

"There remains but one course for the recovery of a sound and healthy condition," the English philosopher Francis Bacon wrote in his *Novum Organum Scientiarum,** or *The New Instrument of Science,* of 1620, "namely, that the entire work of the understanding be commenced afresh." Reject syllogisms as the foundation of investigations into nature; they can rest on assumptions that are invalid. Instead, let "the mind itself be from the very outset not left to take its own course, but guided at every step; and the business be done as if by machinery." On the Continent, Descartes advocated a similar approach in his 1637 *Discourse on the Method.* His method was to strip away every possible observation and interpretation, all of which would be subject to human biases, and to keep stripping away until all that remained was incontrovertibly, objectively true. That truth would be your own existence, which you could prove† because you perceive yourself: *Cogito, ergo sum.*

As first feints go, these tracts were accurate enough: If you take seriously the study of the universe large and small, you have to chal-

---

* The old *Organon* being the collective title that Aristotle's earliest disciples bestowed upon his writings on logic.

† Or at least prove to your own satisfaction. The rub: "A Chinaman of the T'ang Dynasty—and, by which definition, a philosopher—dreamed he was a butterfly, and from that moment he was never quite sure that he was not a butterfly dreaming it was a Chinese philosopher. Envy him; in his two-fold security." Tom Stoppard, *Rosencrantz and Guildenstern Are Dead.*

lenge your notoriously unreliable senses until you know you can trust the evidence, and you have to exhaust logical possibilities until only one explanation remains. In that sense, Bacon and Descartes were being prescriptive: *Here's what you do if you want to be a New Philosopher.*

But they were also being descriptive. They were trying to codify a revolution in thought that was already under way: *Here's what they're doing, these New Philosophers. And here's why you should trust their results, as contrary to received wisdom as they may be.*

In his *Astronomia nova* Kepler led his reader laboriously through his observations and calculations. For more than six hundred pages, he led his readers. He even included his wrong turns, as if the admission of conclusions he rejected fortified the conclusions he embraced. Probably they did. Kepler was letting his readers in on a little secret: *I get stuff wrong.* In doing so, he was letting them in on a bigger revelation: *Here's how I know when I've got stuff right.* In one extreme case Kepler showed not his acceptance of a wrong conclusion but his *rejection* of a *correct* solution. "What a silly bird," he wrote, describing the moment he realized, some seven years into his calculations, that he had reached the same formula several years earlier but had rejected its implications back then because it showed that the orbits of planets might be elliptical, not circular.

Galileo followed the same first-person narrative model in *Sidereus Nuncius*. Like Kepler, Galileo led his readers methodically through his discoveries. First he described his new instrument and how it worked; then he included drawings of his night-by-night, then hour-by-hour, observations. Anyone who followed along could repeat the process and — if they could build or buy a telescope of their own — see the evidence for themselves.

Kepler and Galileo and other New Philosophers were doing what comes naturally, at least to the human brain: They were telling a story in such a way that their readers would understand — so that the narrative would make sense to readers in the same way it made sense to them. Now it was Newton's turn.

From his own perspective, he'd already followed that pattern in the first edition of the *Principia*. In Books One and Two, both called *On the Motion of Bodies,* Newton had presented observations (*Do you see what I see?*) from which, in Book Three, *The Motion of the World,* he derived the universal law of gravitation (*Do you follow my reasoning?*). As a New Philosopher in good standing, as well as one human being trying to get his ideas across to another, he'd done what he was supposed to do in order to persuade readers to trust his results.

Yet he hadn't been persuasive enough. The second edition of the *Principia* presented both Cotes and Newton with the opportunity to try again — to make sure they presented sufficient evidence and logic to convince readers that gravitation was real and, just as important, that the New Philosophy worked. To do so, they would follow the narrative example of the New Philosophy itself: Cotes will tell the reader what Newton will be doing; then Newton will do it; finally, Newton will tell the reader what he's done.

In his editor's preface Cotes introduced readers to the third stage in the development of the study of the motions of matter, the ascendance of scholars "whose natural philosophy is based on experiment." These New Philosophers, he said, proceed "by a twofold method." The first is "analytic": "From certain selected phenomena they deduce by analysis the forces of nature and the simpler laws of those forces." (*Do you see what I see? Do you follow my reasoning?*) The second part of the twofold method, he promised, would be "synthetic" — the generalization of the laws to other phenomena.

Now that Cotes had told the reader what Newton would be doing, Newton did it. At the beginning of the new version of Book Three, Newton introduced his "Rules for the Study of Natural Philosophy." These four rules were in fact a reduction and partial rephrasing of the first edition's nine "Hypotheses" — a vague word, one that confused readers and rightly drew criticism as to whether he was adhering to the New Philosophy. Now, though, Newton would try to formalize the reasoning he would ask his readers to follow.

The first rule held that "Nature is simple." If we could trace one effect to a cause, we could stop right there. Nature, Newton said, "does not indulge in the luxury of superfluous causes."

The second rule expanded on the first. It held that if we could trace one effect to one cause, then we could trace similar effects to the same cause. Newton gave some examples: the cause of "respiration in man and beast," of "the falling of stones in Europe and America," of "the light of a kitchen fire and the sun."

The third rule, in turn, expanded on the second. If you found a principle that applied to specific circumstances, you could apply it more generally to other sufficiently similar circumstances.

Newton was careful to frame the rule within the boundaries

of experiments. He wasn't saying that assumptions should be generalized, only verifiable phenomena. For instance, "the extension, hardness, impenetrability, mobility, and force of inertia of the whole arise from the extension, hardness, impenetrability, mobility, and force of inertia of each of the parts"—the "parts" having already been identified as "the undivided particles of body"—"and thus we conclude that every one of the least parts of all bodies is extended, hard, impenetrable, movable, and endowed with a force of inertia." If we erase the distinction between *over here* and *over there*—between Europe and America, for instance—shouldn't we also erase the distinction between *down here* and *up there,* Earth and heavens? If the cause of the falling of stones is the same in Europe as in America, then the cause of the falling of stones is the same on Earth as anywhere else in the universe—a universe we now have every reason to suspect is full of stones:

> If it universally appears, by experiments and astronomical observations, that all bodies about the earth gravitate toward the earth, and that in proportion to the quantity of matter which they severally contain; that the moon likewise, according to the quantity of its matter, gravitates toward the earth; that, on the other hand, our sea gravitates towards the moon; and all the planets one towards another; and the comets in like manner towards the sun; we must, in consequence of this rule, universally allow that all bodies whatsoever are endowed with a principle of mutual gravitation.

Now came the final part that Cotes promised, the one where Newton would tell the reader what he'd done. Or, more to the point, he would tell the reader what he'd *not* done, and why he'd not done it.

The prevailing wisdom — the philosophical method in general, and certainly as Descartes framed it — asked a reasonable question: How do you take something that makes no sense — a phenomenon that, at least at this point in history, involves none of the senses — and explain it? The prevailing wisdom even supplied a reasonable answer: *You can't.*

Newton, however, wanted to present an alternative. Or maybe he *had* to present an alternative answer to the question of how to take something that makes no sense and explain it.

The answer wasn't: *You can't.*

It was: *You don't.*

Or, perhaps more accurately: *You don't* because *you can't.*

You don't provide a cause for the effects of gravity because you *can't* provide a cause for the effects of gravity, at least according to the current understanding of the workings of the universe. So rather than flail and conjecture and guess and strain to describe it, Newton wrote, let's just take what we do know: the match between math and motions.

"I have not as yet been able to deduce from phenomena the reason for these properties of gravity, and I do not feign hypotheses," Newton wrote toward the end of the General Scholium. "For whatever is not deduced from the phenomena must be called a hypothesis; and hypotheses, whether metaphysical or physical, or based on occult qualities, or mechanical, have no place in experimental philosophy."

By "I do not feign hypotheses" — *Hypotheses non fingo* in the original Latin — Newton wasn't just disavowing the need to identify how gravity works, though he certainly was doing that. He was also articulating a philosophical transition that was already under way, one that would reverse the traditional — the natural, the rational — process of thinking about causes and effects.

We think of causes as preceding effects. And they do. Let go of a clot of earth — cause — and the clot of earth will fall — effect. Descartes, for instance, had assumed that the New Philosophy's investigations of nature would follow that same progression: Identify causes; find effects. His "vortex" theory was a fine example of this model. Descartes knew the rotational motions of the planets, thanks to Kepler's careful bookkeeping. And he could assume that something must be moving them in their rotational orbits. Then he hypothesized: The something that must be moving them in their rotational orbits must itself be moving in some sort of rotational motion — a vortex, if you like. Then he sought to describe the motions of these vortices — the causes — that would create the actual behavior of the planets — the effects.

But in the alternative methodology, effects precede causes: not that a clot of earth falls before the hand releases it, but that the *study of* the effect of falling precedes the possible *identification of* the cause. Or, as in the case of gravity, the *non*-identification of the cause.

On the basis of this reasoning, Newton concluded that we do not need to know how gravity works. We know that it does work, because: We can see its effects; we can derive its math; we can arrive at laws that we can generalize across the universe. We know it's there.

"It is enough," he wrote, "that gravity really exists and acts according to the laws that we have set forth."

◈

It was not enough.

Newton was still asking a lot of readers, especially those of a philosophical bent. Just because Galileo had insisted that the book of nature is written in the language of mathematics didn't mean that *all* of it was. One such objection surfaced in 1719, six years after the publication of the second edition of the *Principia*. An amateur philosopher by the name of George Gordon began his *Remarks upon the Newtonian Philosophy* with a reminder that ignorance of the cause of motions was hardly new — and hardly a promising foundation for a new system of philosophical thought. His book opened: "There is nothing to which the Philosophers of all Ages have so generally applied themselves, as to the Study of Nature; and there is no Study in which they have been able to make so little Progress as in this, especially in that Part of it which is employ'd in discovering the Causes of the beautiful Order and steddy Motions of the Sun, Moon and Stars." These philosophically fanciful structures, Gordon argued, were all "built either upon gross Absurdities, or Words without Meaning." The present Newtonian edifice, alas, was no exception: "I confess that the many Instances of unintelligible Causes that have been assigned by the great Philosophers for the Production of those surprising Phaenomena of the Motions of the Heavens make me suspect this Cause, which looks as monstrous as any of the Fictions of Antiquity, and the Mathematical Dress of the Arguments which support that Cause, does not hinder me from suspecting their Sufficiency."

The Scottish mathematician George Pirrie found Gordon's book so objectionable that he responded with a book of his own. In the introduction he recounted the trajectory of his indignation: First he couldn't believe that someone as obtuse as Gordon was even real, so he had resolved to ignore the book; when Gordon's book reached its second printing, Pirrie could contain himself no longer. *A Short Treatise of the General Laws of Motion and Centripetal Forces: Wherein, By the by, Mr. Gordon's Remarks on the Newtonian Philosophy are, in a few Corollaries and Scholies, Clearly confuted* was primarily a refutation of Gordon's mathematics, but in his opening remarks Pirrie also made sure to put the discussion into its proper historical context: *Real* philosophers now regard mathematical equations as a reliable means of generalizing a description for the motions of matter.

They capture past motions. They account for current motions. They predict future motions.

The prediction of future motions was in fact a component of the New Philosophical method that previous philosophical regimens had lacked. That you could examine existing data and derive a mathematical correspondence rigorous enough that you could elevate it to a law wasn't a definitive argument in favor of the law. The New Philosophy therefore carried a further stipulation: Conduct an experiment. Make a prediction and test it.

One such test had already occurred, though its possible importance regarding gravitation was evident only in retrospect. In 1672 the French astronomer Jean Richer performed various observations near the equator, including mapping the southern sky. Among the equipment Richer packed were several models of a "seconds pendulum"—a pendulum that precisely counts one second with each complete swing, downward and upward.

Tick-tock: one second.

Tick-tock: one second.

Tick-tock: one second.

And so on: 86,400 tick-tocks every day.

Richer found, however, that the pendulums that worked meticulously in Paris were unreliable at the equator—or at least reliably unreliable. The same pendulums that counted 86,400 seconds per day in Paris were beating at the rate of 86,252 at the equator—two minutes and twenty-eight seconds short. They were running slow.

Richer could adjust, and he did. The period of a pendulum swing—the full motion back and forth—depends on its length: the longer the pendulum, the slower its motion; the shorter the pendulum, the faster its motion. Richer needed to speed up his pendulums, so he shaved each three-foot shaft, a bit at a time, until its period matched local clocks (as well as measurements of stars). But neither he nor—upon his return—his fellow philosophers in Paris could explain the difference.

Some fifteen years later, Newton thought he could. When he was composing the *Principia*, he turned to Richer's measurements to justify a claim he was making: A spinning globule of (mostly) water on its surface, such as Earth, should be flatter near the poles and more bulbous at the equator. This bulbousness, he argued, would be due, as the motions of all matter are due, to a combination of a forward motion and a downward motion. On a spinning Earth, the forward motion would be strongest at the equator, while the downward motion, because of the distance from the center, would be weakest there. Things still wouldn't be flying off the surface of the Earth; the downward motion would still be greater than the forward motion. But the balance would have shifted in the forward motion's favor, at the expense of the downward motion.

The downswing and the upswing, the give and the take, the *tick* and the *tock,* depend on the pendulum's weight. *Weight,* though, is just another way of saying *gravitational effect.* The weaker the downward effect of gravity, the freer the motions of the pendulum and the faster the measurement of time.

Newton used Richer's results — the measurements of the pendulums at the equator and in Paris, the difference in the latitudes of the two locations, the predicted bulk of the bulge — in the *Principia* as an after-the-fact validation of his inverse-square mathematics. He cited similar pendulum measurements. In 1677, on an expedition to Saint Helena, an island in the South Atlantic a thousand miles off the African coast, Halley found that his pendulum seemed to *tick* and *tock* more slowly than in London (though Halley didn't keep exact records). A 1682 expedition to the island of Gorée, just off the western bulge of Africa, and the Caribbean islands of Gaudeloupe and Martinique, roughly where the Caribbean Sea meets the South Atlantic Ocean; further expeditions over the years to Paraíba (eastern bulge of South America), Cayenne (northeast South America), Grenada, Saint Kitts, Santo Domingo, Portobello (all in the Caribbean): The results were consistent — the pendulums always swung more slowly near the equator. For Newton, however, these reports offered only supporting evidence, albeit enough to advance a compelling argument. The measurements of Richer, though, were meticulous enough that they were all Newton needed.

Or so he felt. Others felt differently, in three distinct ways.

Under the experimental model, one reasonably certain result doesn't preclude an alternative interpretation. Huygens provided one only three years after the publication of the *Principia*. Huygens had independently reasoned that the greater rotational rate at the equator would cause flattening at the poles and bulging at the equator, in agreement with Newton, but he argued that no additional cause — a cause leading to the downward, centripetal effect — was necessary to account for the seemingly odd behavior of Richer's pendulums. The rotational rate could account for *all* the flattening and *all* the bulging.

Critics could also point out that Richer's measurements didn't meet the rigorous demands of the experimental method. They weren't even the result of an experiment. Richer hadn't designed the setting and the variables in an attempt to discern the shape of the Earth or that shape's inverse-square influence on the period of a pendulum.

Finally, the investigatory method behind the New Philosophy demanded not just affirmation but reaffirmation. Even if Richer had designed and executed an experiment with unprecedented precision, it would still be only one result. Just as discovering a correspondence between existing evidence and new equations wasn't a definitive argument in favor of a law, neither was conducting a single experiment that affirmed a prediction. The experimental method therefore had two further requirements: that the experiment be replicable, and that the experiment be replicated. At least one more test would be necessary before anyone could make a persuasive yay-or-nay argument regarding gravity and the shape of the Earth.

Even as the second edition of the *Principia* was going to press, in 1713, that test was under way.

In 1701, a team of French explorers had completed a geographical survey stretching from Paris virtually due south to the Pyrenees. The purpose of the expedition was practical: to determine with precision the locations of latitudes, information that would be indispensable in matters of trade and war. A decade later the French government authorized another geodesic survey, this time from Paris virtually due north to Dunkirk. The purpose of this expedition was also practical, but it was ideological as well: What is the shape of the Earth?

Did it bulge at the equator, or did it not?

If the Earth rotated within empty space, as Newton held, it would widen at the equator and flatten at the poles. If the Earth rotated within a Cartesian vortex — which would, in effect, serve as a corset — it would slim down at the equator but bulge at the poles. The stakes in this rivalry were not low. The world may have been running out of lands to discover, to conquer, to colonize, but the New Philosophy was opening new horizons of the mind. In the end, England or the Continent would be able to claim the eternal glory of having met the ancient challenge of defining the motions of matter.

To the French, the English were amateurs, enthusiasts, dull in intellect and dull in affect. When the finance minister Jean-Baptiste Colbert founded the Académie des Sciences in 1666 with significant royal investments, he constructed a new observatory near Paris and hired two of the leading astronomers on the Continent —

the Dutch Christiaan Huygens and the Italian Giovanni Domenico Cassini. The French approach to the New Philosophy, he as good as declared, would rely on expertise and technology.

To the English, the French (along with their hired hands) were, well, not English. Worse, the French didn't hesitate to appropriate what *was* English: The English had founded the Royal Society in 1660; the founding of the Académie des Sciences followed six years later. The publication of the *History of the Royal Society*, perhaps not coincidentally, followed the founding of the Académie by one year. It summarized, at some length, the natural superiority of the society's homeland: "England may justly lay claim to be the head of a philosophical league above all other countries in Europe." Among the many unique virtues of England and the English that the *History* cited were: "an unaffected sincerity"; "an honorable integrity"; a "love to deliver their minds with a sound simplicity"; a reserve that makes them "not extremely prone to speak"; a concern for "what others will think of the strength [rather] than of the fineness of what they say"; "a neglect of circumstances and flourishes"; "a genius so well proportioned for the receiving and retaining its mysteries"; "the position of our climate, the air, the influence of the heaven, the composition of the English blood, as well as the embraces of the ocean," all of which "join with the labors of the Royal Society to render our country a land of experimental knowledge." Oh, and a "universal modesty."

The Earth rotates on an axis, and we call the extremes of that axis the North Pole and the South Pole. If you're looking at the stars in the night sky, the rotation of the Earth will give the illusion that the stars are rotating around a single point. In the Northern Hemisphere, that point will be due north of the North Pole—and for that reason we call the star that happens to occupy that position the Pole Star or the North Star. If you were at the North Pole, the star would be directly overhead. At that location, the star and the surface of the Earth would describe a right angle, or 90 degrees, with you as the point where the vertical and horizontal legs of the right triangle meet.

Then again, if you were at the equator, the Pole Star would appear to rest on the horizon, or at zero degrees relative to you. If you were standing anywhere between those two points, you could determine your latitude in the same way: Measure the angle between the North Star and the horizon: 1 degree, 2 degrees, 3 degrees, 18 degrees, 47 degrees, up to 90 degrees.

If the surface of the Earth flattens somewhat near the poles, as Newton predicted, then the land surface between those latitudes would lengthen as you head northward. For instance, if you were traveling between the point on the Earth that describes a 70-degree angle with the Pole Star and the point on the Earth that describes a 72-degree angle with the Pole Star, the distance between 70 degrees and 71 degrees latitude would not be as long as the distance between 71 and 72.*

According to Descartes, though, the Earth bulges near the poles, meaning that the land surface between latitudes would shorten as you head northward.

Giovanni Domenico Cassini oversaw the first survey, stretching

---

* The same principles apply in the Southern Hemisphere, though no visible-to-the-eye star occupies the position above the Earth's axis there.

from Paris south to the Pyrenees; on its own, it wasn't revealing one way or the other. After Giovanni's death in 1712, his son Jacques Cassini assumed the directorship of the Paris Observatory as well as the organization of the second expedition, stretching from Paris north to Dunkirk. Jacques hoped this second survey might cover enough literal ground that the combination of the two surveys would reveal the difference in distance between one latitude and the next. And it did, at least to the satisfaction of Jacques: A degree of latitude at Dunkirk was about nine hundred feet shorter than a degree of latitude in the Pyrenees.

Tie score: Newton/Richer 1, Descartes/Cassinis 1.

"A Frenchman who arrives in London," wrote a Frenchman who arrived in London in 1726, "will find philosophy, like everything else, very much changed there." François-Marie Arouet, writing under the name Voltaire, listed a few of the most salient differences. On the Continent, space was full of some sort of matter; in England, it was a vacuum. On the Continent, the Moon pressured the tides; in England, the Moon attracted the tides (and vice versa). On the Continent, "light exists in the air"; in England, "it comes from the sun in six minutes and a half." In all these cases, the difference depended on the interpretation of what moves matter. In France, an invisible physical matter exerted its influence; in England, an invisible physical *something* exerted its influence. It might be matter. It might not.

Voltaire had charged himself with the challenge of explaining Newton's work to his fellow countrymen. He began with a pretense of equanimity. "According to your Cartesians," Voltaire wrote, "everything is performed by an impulsion" — the effect of contact with vortices — "of which we have very little notion; and according to Sir Isaac

Newton, it is by an attraction, the cause of which is as much unknown to us." The mystery behind the motions — their cause — would seem to render both interpretations nonsensical — or, perhaps, nonsensible, in that they didn't rely on the evidence of the senses.

Attraction, Voltaire acknowledged, is a slippery concept. "Sir Isaac Newton, after having demonstrated the existence of this principle, plainly foresaw that its very name would offend; and, therefore, this philosopher, in more places than one of his books, gives the reader some caution about it." Attraction implied active participation — the act of attracting — which was more than Newton was willing to grant. Newton wanted to acknowledge only the existence of effects between bodies — effects that one can observe — and a mathematical relationship between the bodies that matched the effects. Newton "bids [his reader] beware of confounding this name with what the ancients called occult qualities." Instead, as Voltaire explained, Newton was asking readers to trust him and his methodology — "to be satisfied with knowing that there is in all bodies a central force, which acts to the utmost limits of the universe, according to the invariable laws of mechanics."

Force. If Voltaire was going to add caveats to Newton's use of the word *attraction,* then he should have done the same with *force,* which also implied participation. Not necessarily active participation; perhaps passive participation. But participation nonetheless. Voltaire was learning that in matters of gravitation, some of the concepts were so counterintuitive that description could be problematic, language itself perhaps inadequate. But Voltaire's larger point was Newton's: Follow the logic wherever it may lead; resist the impulse to assign causes in the absence of evidence; accept the math, because *it is enough.*

It still wasn't enough, alas, but it was getting there. As with many aspects of the New Philosophy, Voltaire was the town crier who brought a Newtonian concept—in this case, the effect-and-cause approach—to awareness in France, and therefore on the Continent. "If we see [a cause's] effects," he wrote, "we know it exists; we must *not* begin by imagining the cause and then making hypotheses, because that is the surest way of losing our way. Instead, let us follow step by step what actually happens in Nature." The metaphor he invented was especially evocative for an exploration-savvy populace: Imagine you're a voyager who has discovered the mouth of a river. Do you assume you know the source of the river, or do you travel up the river to discover it for yourself?

"Very few people in England read Descartes, whose works are indeed useless," Voltaire concluded, airily. Thanks in part to Voltaire's proselytizing, both in his 1733 *Letters on the English* and through private conversations and correspondence with the leading philosophers in France, the intellectual balance began to shift. The pendulum, as it were, was swinging. By the time Voltaire's *Elements of Newton's Philosophy* appeared in 1738, a reviewer could proclaim, "All Paris resounds with Newton, all Paris stammers Newton, all Paris studies and learns Newton."

Among the most passionate students, if not stammerers, of Newton in Paris was Pierre Louis Moreau de Maupertuis, not just a mathematician but, more important for the purposes of pursuing a validation of the law of gravitation, the director of the Académie des Sciences. If, as was increasingly likely, the credit for the formulation of the new conception of the universe wouldn't be France's, at least maybe the decisive tiebreaker measurement of the Earth's shape would be.

The two Cassini expeditions had together covered only the length of France, from the Strait of Dover in the north to the Mediterranean Sea in the south. To ensure a higher level of precision, Maupertuis decreed that the new measurements would extend north all the way to the Arctic Circle and south all the way to the equator.

The enterprise was broad and bold, its champions proud. The explorers were nothing less than "the new Argonauts," declared Bernard Le Bovier de Fontenelle, the academy secretary. Fontenelle had lived through a lot. He was old enough to have published his own explanation of the Copernican interpretation of the universe, *Entretiens sur la pluralité des mondes,* or *Conversations on the Plurality of Worlds,* a year before Newton published the *Principia.* And because he had the foresight to write it not in Latin but in French, in order to ensure a popular readership, he had become one of France's foremost literary figures. In the nearly five decades since then Fontenelle had assumed the role of national advocate of the New Philosophy. In that capacity he had also witnessed every embarrassment on the French side of the feud between Newtonians and Cartesians. Now nearly eighty, Fontenelle decided that if the French were going to cede the greatest philosophical battle of the past century, or of all history, they would do so on their own swashbuckling terms. "How many hardships, and fearful hardships, must accompany such an enterprise?" Fontenelle exhorted the academy in 1735. "How many unforeseen perils?"

He had no idea. Tropical diseases, the suspicions of local authorities, the indifference of natives, a crew harboring both apathy toward science and antipathy toward authority: All conspired to make the equatorial expedition to the home base of Cuenca, only 3 degrees south of the equator, a years-long endurance test—a test that the team's surgeon, for one, failed. Having an affair with the daughter of a patient, thereby enraging her relatives and scandalizing the villagers; flogging a *mestizo* without first alerting local authorities; warning a city official that he would "cut off his ears"; responding to a possible insult during a holiday celebration by brandishing a pistol and commanding his slave, "Kill them all!"—a request that was unrealistic, given the size of the mob: Perhaps the surgeon had taken Fontenelle's exhortations too much to heart. In any case, in the ensuing riot he was stoned, piked, and lanced, and somewhere in that bloody cacophony came the opening of the wound that killed him several days later.

No surprise, then, that the results of the equatorial expedition took a while to reach Paris. Not so the data from the northern excursion to Lapland, which Maupertuis himself had elected to join. The expedition was relatively free of hardship, as far as expeditions through extreme physical landscapes go. Ridiculously cold and relentlessly dark in the winter, insanely insect-ridden in the summer —*naturellement*. But the Tornio River offered a route along a nearly north–south axis, ideal for making measurements from one latitude directly north to the next. To the extent that the river diverged from that straight line, the explorers could leave the Tornio and complete their measurements within the valley. Local boatmen helped navigate the thunderous rapids; the native population opened their homes on a regular basis, an especially welcome luxury during the brutal winter of 1736–37; and throughout the darkest months, when "the sun scarcely rose above the horizon"—as Maupertuis wrote in his account of the expedition—"the twilight, the white snow, and the Aurora Borealis supplied enough light for four or five hours of work daily."

The expedition returned to France in 1737. The data from the equator wouldn't arrive for a few more years, but the measurements from the Arctic alone indicated that a degree of latitude was longer in the north than in France.

Newton 2, Descartes 1.

Voltaire sent his congratulations to Maupertuis: "My dear flattener of the world and Cassinis."

Even so, the shape of the Earth involved a Newtonian prediction *down here*. Validation of a Newtonian prediction *up there* would still be necessary in order to justify the *universal* part of the law of universal gravitation.

The French pressed their advantage. True to their conviction that they were more rigorous about natural philosophy than the English, they seized for themselves the next big test of gravitation: the return of a comet on or about 1758.

Test, actually, might be too strong a word. It *was* a test, but it was also an affirmation. The question increasingly wasn't whether the comet would return but when. How accurately could mathematicians and astronomers use Newton's math to predict the perihelion — the point in the comet's orbit when it is nearest to the Sun — using the law of gravitation?

Edmond Halley had first written to Newton about this comet in September 1695: "I am more and more confirmed that we have seen that Comett now three times, since ye Yeare 1531." But in a universe full of mutually gravitating bodies — bodies that over the course of their non-repeating elliptical orbits occupy positions relative to one another that are moment to moment uniquely new — the timing between a comet's visits wouldn't be straightforwardly predictable. The interval between the perihelion of 1531 and the perihelion of 1607 had been over seventy-six years, while the interval between 1607 and 1682 had been less than seventy-five years. As Halley wrote to Newton, "I must entreat you to consider how far a Comets motion may be disturbed by the Centers of Saturn and Jupiter, particularly in its ascent from the Sun, and what difference they may cause in the time of Revolution of a Comett in its so very Elliptick Orb."

Halley became the leading popularizer of the comet's potential for the study of astronomy and, with it, the understanding of gravitation among the public. In 1705 alone, Halley published a paper on comets in the Royal Society's *Philosophical Translations;* a highly mathematical pamphlet, *Astronomiae cometicae synopsis,* in Latin; and, in English, a less formal presentation of that same material, *A Synopsis of the Astronomy of Comets.* He wrote a revised edition of the latter book in 1719, but it didn't appear until 1749, seven years after his death. Having taken into consideration the comet's fairly close approach to Jupiter in 1681, Halley concluded: "It is probable that [the comet's] return will not be until after the period of 76 years or more, about the end of 1758, or the beginning of the next."

The reappearance of Halley's book attended the pending reappearance of the comet. "Comets hav[e] of late been a prevailing
topic of most private as well as public conversations," *Gentleman's
Magazine* reported in January 1756. The publication reprinted passages from Halley's *Synopsis* and Newton's *Principia;* those excerpts
also appeared that same month in the *Gentleman's and London Magazine,* published in Dublin, and the *Newcastle General Magazine.*
The *General Magazine of Arts and Sciences* presented, as part of a
yearlong series on the solar system, a two-part discussion of comets.
An anonymous (for good reason) author published *An Account of
the Remarkable Comet, Whose Appearance Is Expected at the End of
the Present Year 1757, or at the Beginning of 1758.* It was essentially a
work of plagiarism — a compendium of material that had appeared
elsewhere, much of it word for word. It was also a best-seller, going
through three printings in a year.

While the English watched the skies, the French mathematician
and physicist Alexis-Claude Clairaut set himself the task of refining
the prediction for the comet's return. The calculations proved to be
even more difficult than Halley might have imagined. Clairaut assumed that he would have to take into account the effect of Jupiter
and Saturn only when the comet passed close. But no, he soon realized: He would have to calculate Jupiter's and Saturn's contributions over the entire orbit of the comet. No, make that two orbits:
He would also have to factor in Jupiter's interaction with the Sun,
which required comparisons over two of the comet's periods, which
is to say, over a stretch of 150 years. Actually, make that *three* orbits;
he would need a third to test the accuracy of his methods. In the
end Clairaut had to calculate the positions of the comet and the two

giant planets, then compare his calculations to the actual observations in the astronomical record, more than seven hundred times.

He announced his results to the Paris academy in November 1758. First, Clairaut reported that his mathematical methods had allowed him to predict (retroactively) the timing of the comet's three previous visits, in 1531, 1607, and 1682. According to his math, the period between the perihelion of 1531 and that of 1607 "should" have been 432 days longer than the period between the perihelion of 1607 and that of 1682. Then he had checked his math against the motions — actual astronomical observations from the sixteenth and seventeenth centuries. They showed that the difference was not 432 days but 459 days.

The difference between prediction and observation — between math and motion — was only 27 days over three-quarters of a century: a match at a 99.9 percent level of accuracy. Incorporating a similar margin of error, and acknowledging that other bodies (for instance, an unknown planet farther out in the solar system) might also exert an influence on the comet's trajectory, Clairaut predicted that the comet would reach perihelion about mid-April 1759, give or take a month.

The comet reached perihelion on March 13.

The French could claim credit for the accuracy of the prediction. But the English — deep breath to appreciate the virtues of imbibing a climate on an island embraced by an ocean — could claim Newton. As Halley had written in his *Synopsis:* "Wherefore, if agreeable to my prediction, [the comet] shall return again about the year 1758, impartial posterity will not scruple to ascribe this invention

to an Englishman." In November 1759, a pseudonymous colum-
nist wrote in *Gentleman's Magazine:* "I cannot but congratulate my
countrymen on an event so glorious to the Newtonian doctrine of
gravity, and to the memory of that excellent philosopher Dr Halley;
and may it ever be remembered that the first instance of an event of
this kind was foretold, and with accuracy too, by an Englishman."

National pride aside, the English *General Magazine of Arts and
Sciences* summarized the reigning sentiment among laypeople and
scholars alike, in England and abroad: "It cannot but excite the Attention and Admiration of the Curious in general, and fill the Minds
of all Astronomers with a ravishing Satisfaction, as it has, by this Return, confirmed Sir Isaac Newton's *Rationale* of the Solar System,
verified the Cometarian Theory of Dr. Halley, and is the first Instance of Astronomy brought to Perfection."

Whatever the implications for the cross-Channel rivalry, at least
the British Isles and the European Continent could agree: From now
on, the heavens would be yet another object of investigation. The
previously unapproachable, forever inaccessible realm of all that's
up there would be approachable and accessible from *down here.*

We couldn't actually go up there, as in Kepler's fantasy, but now
we could *get* up there, through the magic of mathematics. No: the
non-magic of mathematics. The logic of mathematics. We could
avail ourselves of the tool that Galileo had celebrated as a major
contribution to, and that Newton had appropriated as an essential
component of, the New Philosophy: the match between math and
motions. The universe obeyed laws, and we could discover them.

In which case: Why *not* us?

GRAVITY AS A FACT

The comet came back.

But that's not all.

The comet came back precisely according to mathematical predictions of greater and greater sophistication.

But that's not all.

The comet's return foretold a celestial cotillion: a dance among the planets; a dance among the stars; a dance among the planets and our own star; and even, all wallflower-like, standing with its back against what would once have been the sphere of the fixed stars but was now just a neutral place in space, a new planet.

But most of all the comet heralded the birth of a new universe to navigate; the beginning of a new understanding of our own place within that whole; the certainty that the universe *is* a whole, a *down here* and an *up there* that we can treat as one; the confidence that this new unification can extend beyond astronomy and physics to other investigations of nature; the faith in ourselves that, having conquered gravity, surely we could conquer anything — whether the clockworks of a Creator or the machinations of man.

No longer need New Philosophers concern themselves with debates over the existence of gravitation. Instead, they might wonder to what extent they could use this newfound knowledge about the motions of matter — and their newfound familiarity with the method that allowed them to gain this knowledge — to change the equation of the universe.

Nature doesn't add up. Take everything we know about the universe, then take everything there is to know about the universe, and the two won't be equal. They never have been equal, and almost certainly they never could be. No matter how much we might manage to learn about the workings of the universe, some mysteries will remain.

We can, however, correct that imbalance. We can make the *what-we-know-about-nature* side of the equation equal to the *what-there-is-to-know-about-nature* side simply by inserting a variable, an x. This x would be equal in value to whatever is absent. Whatever its value, this x would balance the two sides of the equation of the universe. Take everything we know, add an x equal to everything we don't know, and nature adds up after all.

Everything we know + everything we don't know = everything there is to know.

Obvious. Simple. Simplistic, even. Or so it seemed once New Philosophers began to recognize the formula for what it was: the way old philosophers had thought about the world. The old philosophers, of course, hadn't thought in terms of x; Descartes had added that notation to the algebraic vocabulary in his 1637 *La Géométrie*. The New Philosophers didn't necessarily think in terms of x, either, when they measured the knowledge of nature. But measure the knowledge of nature they did, and in a world that more and more was revealing itself to be operating according to mathematical laws, the universe itself was emerging as an equation to solve. From the perspective of the New Philosophy, premodern thinkers had closed the gaps in their knowledge of nature, as often as not, by invoking something beyond nature — beyond the natural, which is to say, the *super*-natural. Here be dragons, in all their guises: god, gods, demigods, demons, the dead — you name it, then swap it in for x.

For natural philosophers — as the New Philosophers had begun calling themselves now that their philosophy was not so new — the return of a comet in 1759 provided the symbolic transition between two ways of regarding the x in the equation of the universe. On one side of the historical moment, the French mathematician Pierre-Simon Laplace wrote half a century later, the world was full of "the fears begotten by the ignorance of the true relationship of man to the universe." On the other, "the learned world awaited with impatience this return which was to confirm one of the greatest discoveries that have been made in the sciences" — the law of gravitation.

"Let us recall," Laplace continued, "that formerly, and at no remote epoch, an unusual rain or an extreme drought, a comet having in train a very long tail, the eclipses, the aurora borealis, and in general all the unusual phenomena were regarded as so many signs of celestial wrath. Heaven was invoked in order to avert their baneful influence." The specific example that Laplace cited was "the long tail of the comet of 1456," which "spread terror throughout Europe." That comet's return in 1682 spread terror throughout Newton and Halley, too, and of a particularly biblical kind. Looking backward in history, Halley presented a paper to the Royal Society in 1694 suggesting that the biblical Flood might have been the result of a comet. Looking forward, Newton assumed that in time the elliptical orbit of the comet would intersect with the focus of the planetary system — the Sun — and that the resulting conflagration would be the fulfillment of biblical prophecy.

Three decades later Jonathan Swift would satirize these sorts of superstitions in *Gulliver's Travels*. The population on the flying island of Laputa "have observed Ninety-three different Comets, and settled their Periods with great Exactness." These advances in knowledge, however, have had a disconcerting effect — "continual Disquietudes," so much so that the first thought of the day for Laputians is "what Hopes they have to avoid the Stroak of the approaching Comet."

By the mid-1750s, fear of the sky falling had coalesced around the return of one comet. In late 1755, the *General Magazine of Arts and Sciences* explained that an encounter between Earth and the comet "would undoubtedly consume the Earth, and all its Inhabitants, as so many Moths; it might convert the Matter of the present Earth into a different Kind of Substance, and render it an Habitation fit for beings of a quite different Nature from ours." In a best-selling pamphlet published in London and other English outposts in

1755 and 1756, the Methodist evangelist John Wesley warned that the comet could reflect divine wrath for a failure to believe sufficiently in God. If so, Wesley promised, it "will set the earth on fire, and burn it to a coal." For an apparent non-empiricist, he was unusually attentive to causation:

> Probably it will be seen first drawing nearer and
> nearer, till it appears as another moon in magnitude,
> though not in colour, being of a deep fiery red; then
> scorching and burning up all the produce of the earth,
> driving away all clouds, and so cutting off the hope or
> possibility of any rain or dew; drying up every foun-
> tain, stream, and river, causing all faces to gather
> blackness, and all men's hearts to fail; then executing
> its grand commission on the globe itself, and causing
> the stars to fall from heaven. O, who may abide when
> this is done? Who will then be able to stand?*

Not that God needs a comet in order to exact his vengeance upon the heathens. Such are the perks of being omnipotent. "As the Almighty created this World without Means," a popular almanac for 1757 clarified, "he can easily dissolve it without any foreknown me-diate Cause."

The question of a Creator — First Cause, as the moderns called it — had always been part of the debate over the New Philosophy. It

---

* Turns out Wesley had confused his comets, using Halley's data from 1680 rather than 1682. The lesson presumably still applied.

was a question that most, if not all, non-Christian renderings of creation could afford to elide, or even to ignore altogether. If a creation myth's pre–*In the beginning* involved a somethingness — a Chaos, a cosmic egg, or some other formless substance — then mythmakers might have to account for a moment of coalescence but not for the creation of the matter itself. The Christian idea of creation *ex nihilo*, however, required a pre–*In the beginning* nothingness. The matter in a Christian universe had to come into existence. If your idea of a universe is one that begins out of nothing, then accounting for the creation of matter is unavoidable, and in the Christian interpretation the assignment of an identity to that First Cause was inevitable: God. To seek "second causes" — to seek the cause of the vibration of a magnet's needle, or of the beating of a heart, or of the fall of a stone, and then to seek the causes of those causes, working one's way backward in time — is to diminish the presence of a Creator in the equation of the universe. Is it not?

It is not, argued the English philosopher Francis Bacon, encountering this argument at the very dawn of the New Philosophy. In his 1605 *Advancement of Learning*, Bacon acknowledged that the pursuit of second causes — of causes in nature, of causes that don't require the *immediate* intercession of God — might indeed lead a lazy mind to dismiss God altogether. But for a more rigorous mind, "a further knowledge brings it back to religion; for on the threshold of philosophy, where second causes appear to absorb the attention, some oblivion of the highest cause may ensue; but when the mind goes deeper, and sees the dependence of causes and the works of Providence, it will easily perceive, according to the mythology of the poets, that the upper link of Nature's chain is fastened to

Jupiter's throne." When several decades later the Royal Society hired the young clergyman Thomas Sprat to write a history of the organization, the purpose was in part to counteract precisely that line of criticism.

A universal theory of gravitation, however, presented a decisive challenge to anti-causists. The discovery that the heart pumps blood throughout the body in a circulatory fashion? Fine, maybe. The discovery that a certain something describes the motions of all matter *down here* and *up there* — indeed, erases the distinction between the terrestrial and the celestial? Not so fine.

In England, one philosophical movement in particular captured the anti-Newtonian sentiment. The theological writer and amateur geologist John Hutchinson wasn't particularly influential during his lifetime, but the posthumous publication of his collected works in 1748 inspired intellectuals who worried that Newton in particular and natural philosophy in general had abandoned First Cause. Hutchinsonianism, as they called it, lasted for decades; at Oxford it was more popular than Methodism.

The core principle of Hutchinsonianism was a familiar one: that Newton had gotten the investigation of the universe backward by starting with effects and seeking causes. Hutchinson didn't just advocate reversing that process. He didn't just want investigations of nature to begin, Descartes-style, with hypothetical causes that lead to effects. Rather, he wanted those investigations to begin with religion's ur-hypothesis, the existence of God: First Cause.

In response to Newton's *Principia* — "the profoundest nonsense ever writ" — Hutchinson published *Moses's Principia*, in which he reinterpreted Genesis as, basically, the story of gravitation. "Gravity had got the better of Revelation," Hutchinson wrote in another posthumously influential book, *Glory or Gravity*. The modern concept of gravity wasn't even necessary; it was, in fact, redundant: In the Old Testament, Hutchinson said, the verb *glory* literally meant "to make heavy."

Hutchinson wouldn't have cared about how to account for the motions of matter after First Cause; he would have been content for God to be all the subsequent second causes, too. But Hutchinson's disciples were a little squeamish on that detail. They liked the Revelation part of his writings. But they worried that if you relied only on Revelation, as Hutchinson advised, you risked resorting to supernatural explanations, and only supernatural explanations, for every subsequent development. Surely, Hutchinson's followers hoped, second causes could include *some* mechanical processes — anything tangible that might ground gravity in a matter-moving-matter way — while also preserving biblical authority.

They got their wish with the release of some of Newton's correspondence in 1744 and 1756. The letters contained all sorts of speculations on Newton's part regarding the physical properties of gravity — but they did so only because they dated to the maybe-gravity-is-"gummy" 1670s, a decade before Newton published the purely mathematical arguments of the *Principia*. Only adding to the Hutchinsonians' optimism was that Newton never entirely abandoned the possibility of a mechanical second cause; in his *Opticks* of 1705 he speculated on the existence of "agents in nature able to make the particles of bodies stick together by very strong attraction."

"He who says that there is a power, motion, or gravity which is not from Him and at His command," Hutchinson had written, "does he not set up another god?" Maybe. But Newton, Hutchinsonians could now answer, to their relief, did not.

What Hutchinsonianism was to theological objections, Romanticism was to aesthetic objections. Romanticism—a loose movement of European artists and intellectuals that rose to prominence in the late eighteenth century and, like Hutchinsonianism before it, endured for several decades—valued nature itself over natural philosophy. Romantics praised the wonders of the world and decried the parsing of its parts.

Newton was, inevitably, one of their favorite targets. A monotype that the English poet and artist William Blake titled simply *Newton,* from the turn of the nineteenth century, depicted the philosopher reaching as far *down here* as possible: perching on a boulder at the bottom of the ocean in the company of algae and anemones, bending forward, pinning a scroll to the seabed with a geometer's compass. Newton, deep in mathematical thought, has lost literal sight of the soaring glories of the universe. Nowhere in the image is a hint of *up there*—which in this interpretation would have to include the surface of both the ocean and the continents, as well as the heights of the heavens. "The souls of 500 Sir Isaac Newtons," the poet Samuel Coleridge wrote, "would go to the making up of a Shakespeare or Milton." *

For the Romantics, though, Newton was just the beginning,

---

* That said, Milton's own attempt at a Dante-esque epic about heaven and hell, *Paradise Lost*, reverted to Aristotelian cosmology even though he wrote it in the 1660s, more than half a century after Galileo's decisive discovery of evidence for the Copernican model.

a symbol of all of natural philosophy and its method of working "backward" from effects to seek cause after cause after cause, sacrificing mystery to math at every step. Even if natural philosophers were to confine themselves to second causes — even if their investigations didn't reach, let alone negate, First Cause — they could do a lot of damage along the way. "Do not all charms fly/At the mere touch of cold philosophy?" the English poet John Keats wrote.

> *Philosophy will clip an Angel's wings,*
> *Conquer all mysteries by rule and line,*
> *Empty the haunted air, and gnomed mine —*
> *Unweave a rainbow*

Or as Blake wrote, with ostensibly damning simplicity, "Art is the tree of life. Science is the tree of death."

Certainly Newton wouldn't have recognized himself in his critics' descriptions. Like Bacon, Newton had come not to bury God but to praise him. "When I wrote my treatise about our Systeme," Newton told the cleric Richard Bentley in their exchange of letters during the winter of 1692–93, "I had an eye upon such Principles as might work with considering men for the *beliefe* of a Deity." Bentley had accepted an invitation to give the first in a series of lectures that the chemist Robert Boyle, himself the discoverer of a law of nature (a correspondence between volume and pressure in gas), had endowed for the purpose of exploring the relationship between Christianity and natural philosophy; Bentley was writing to Newton to see if the universal law of gravitation left God out of the new equation of the universe. It did not, Newton answered, and he cited some of the more peculiar effects of gravitation—for instance, how "all the Particles in an infinite Space should be so accurately poised one among another, as to stand still in a perfect Equilibrium. For I reccon this as hard as to make not one Needle only, but an infinite number of them (so many as there are Particles in an infinite Space) stand accurately poised upon their Points." Newton, keeper of two notebooks, agreed with Bentley that God must intervene in order for the universe to maintain its stability. His argument carried echoes of Philoponus's idea that God imparted impetus to matter at the moment of creation: "Gravity may put the planets into motion, but without the divine Power it could never put them into such a circulating Motion as they have about the Sun."

Twenty years later, in the General Scholium concluding the second edition of the *Principia,* Newton was still issuing reassurances that gravitation poses no threat to a belief in divine intervention. Gravity can explain the *ongoing* motions of bodies in space, Newton argued. It can offer a prediction of future positions of bodies relative to one another. It can offer a postdiction of past positions of bodies relative to one another. These bodies, he wrote, "will indeed persevere in their orbits by the laws of gravity, but they certainly could not originally have acquired the regular position of the orbits by these laws."

For the source of such an exquisite equilibrium, Newton saw only one possibility. "This most elegant system of the sun, planets, and comets could not have arisen without the design and dominion of an intelligent and powerful being," he wrote. "He rules all things, not as the world soul but as the lord of all. And because of his dominion he is called Lord God *Pantokrator.*" (In a footnote, Newton explained: "That is, universal ruler.") "He is eternal and infinite, omnipotent and omniscient, that is, he endures from eternity to eternity, and he is present from infinity to infinity."

Newton wasn't saying that gravity *is* God. Just as God "is not eternity and infinity, but eternal and infinite," and just as God "is not duration and space, but he endures and is present," so God is not gravity. We may deduce causes from effects, he wrote in his *Opticks,* but "the very first Cause," he emphasized, "certainly is not mechanical."

Likewise the natural philosophers who followed Newton: They also wouldn't have recognized themselves in their critics' descriptions. They were searching for second causes not because they didn't believe in God or appreciate mystery. Contrary to what their critics charged, the driving question for natural philosophers wasn't *How much can we ignore First Cause?* Natural philosophers either accepted the existence of God as First Cause — most did, overwhelmingly — or thought the question was irrelevant for the purposes at hand, or both.

Instead, they were consciously adopting the methodology that had led Newton to his laws. For natural philosophers the driving question was *How far can we extend the investigation of second causes without appealing to First Cause?* How far could they follow Newton's gravitational model — Voltaire's effect-and-cause river — to its source?

Probably no mathematician appreciated the descriptive and predictive powers of the law of gravitation as much as Laplace did. Newton had invented calculus in order to arrive at his laws, but he had felt the need to present his math to the public and his peers in the familiar form of geometry. Laplace, however, converted Newton's work back into calculus, and then, in his *Traité de mécanique celeste*, which he published in five volumes between 1798 and 1825, applied those equations to account for the gravitational effects among all known bodies in the solar system. When his former pupil Napoleon asked Laplace where God entered into these all-encompassing equations, Laplace replied, "Je n'avais pas besoin de cette hypothèse" — "I had no need of that hypothesis."

Out of context the response sounds cheeky, and given Laplace's reputation for an unseemly level of self-regard, it probably was to some extent. It also sounds like a First Causist's worst nightmare. But in context the comment acquires a whole other meaning. It arose from Laplace's appreciation of the historical moment — the *philosophical* moment — that he and his contemporaries occupied. Not only had the law of gravitation given them a new means for studying the motions of matter *down here* and *up there*, but *how* Newton arrived at the law had given them a new way to think about the equation of nature. As Laplace wrote, one could now "arrive without the assistance of any hypothesis, but by strict geometrical reasoning, at the principle of universal gravitation."

Let $x = x$. Allow uncertainty to linger until you could discover — or not! — the missing element of the natural world that would explain whatever phenomenon was under examination. Look at the logic behind the law of gravitation and say, as Newton had wished in his General Scholium, "It is enough."

❖

The return of a comet wasn't Halley's only great prediction. The other was less specific, and less specifically astronomical or mathematical, but just as prescient. If the comet did return according to schedule, he wrote in his 1705 *Synopsis,* then astronomers would have "a large Field to exercise themselves in for many Ages, before they will be able to know the Number of these many and great Bodies revolving about the common Center of the Sun and reduce their Motions to certain Rules."

Halley was referring to comets, and after the 1759 return of the one that would bear his name, comets did indeed become their own field of study. The following year Clairaut improved the mathematical analysis of comets, revising his calculations so that they retroactively shaved his thirty-day margin of error regarding the return of Halley's comet by ten days. On the observational side, improvements in telescopic technology over the next few decades allowed, for example, the German-English astronomer Caroline Herschel to discover eight comets between 1786 and 1797.

The "large Field" of discovery, however, didn't stop at comets. It extended into astronomy in general. "So far from having to fear that new observations will disprove this theory," Laplace wrote in his *System of the World,* "we may be assured before-hand, that they will only confirm it more and more." In the 1780s and 1790s Caroline Herschel's brother William, a constructor without parallel of larger and larger telescopes (and, in 1781, the discoverer of the planet Uranus), observed that some individual "stars" were actually two stars. The existence of systems in which two stars orbited each other provided a strong argument that the law of gravity operated at least as far as the stars, which still held the same status as they had in ancient times: the farthest extent of the universe.

Nor did the "large Field" stop at astronomical observations that validated the theory of gravitation. It extended to astronomical predictions. "There still remains numerous discoveries to be made in our system," Laplace wrote. "The planet Uranus and its satellites, but lately known to us, leaves room to suspect the existence of other planets, hitherto undiscovered." The existence of one other planet, anyway. The French mathematician Urbain-Jean-Joseph Le Verrier used irregularities in Uranus's orbital path—departures from Newtonian predictions based on the available evidence—to locate, on paper, the position of a perturbing planet, heretofore invisible. He sent a letter to the German astronomer Johann Gottfried Galle telling him where to look, and the very day the letter arrived, September 23, 1846, Galle and his student Heinrich d'Arrest looked there and found it: the new planet Neptune.

The "large Field" didn't even stop *up there*. In 1797, the British physicist Henry Cavendish demonstrated that gravitation extended *down here*, too—not just in the sense of tides interacting with the Moon or Earth interacting with the Sun, but in the sense of objects on Earth interacting with objects on Earth. Cavendish used a torsion balance—an instrument that guaranteed an exquisite level of stability—to show that in isolation, in the absence of any outside influence, two large lead spheres and two small lead balls "gravitated" toward one another.

At some point natural philosophers began to suspect that they weren't asking the right question. It wasn't *Where else does the "large Field" of gravitational discovery extend to?* It was *Where doesn't it?*

The universe is logical, not capricious. Nature obeys laws, and those laws are universal. "The same principles by which we measure an inch, or an acre of ground," the American patriot and pamphleteer Thomas Paine wrote, "will measure to millions in extent. A circle of an inch diameter has the same geometrical properties as a circle that would circumscribe the universe. The same properties of a triangle that will demonstrate upon paper the course of a ship, will do it on the ocean; and when applied to what are called the heavenly bodies, will ascertain to a minute the time of an eclipse, though these bodies are millions of miles from us."

And if universal laws govern the material parts of nature, maybe they extend to the immaterial parts as well — to human nature, to society. The British philosopher John Locke raised this possibility only three years after the publication of the *Principia:* "The state of nature has a law of nature to govern it, which obliges every one: and reason, which is that law, teaches all mankind, who will but consult it, that being all equal and independent, no one ought to harm another in his life, health, liberty, or possessions," nor in, as Locke mentioned in similar contexts, "the pursuit of happiness." The "law of nature" that Locke cites is the one that future generations of philosophers, revolutionaries, and plain old provocateurs would choose to describe their own historical moment: the Age of Reason.

In *The Age of Reason,* published in 1794, Paine was still invoking Bacon's argument that the study of nature doesn't challenge God but reveals God in all his glory: "The Creator of man is the Creator of science; and it is through that medium that man can see God, as it were, face to face." And the foundational principle was still a rejection of the supernatural — literally so, in the case of Thomas Jefferson's

The Philosophy of Jesus of Nazareth, which consisted of the text of the New Testament minus the miracles. But natural philosophers now understood how far the lessons of gravitation might extend: The laws of nature were not just universal but knowable, and not just knowable but knowable by all.

The collection and dissemination of knowledge — "the art of instruction and enlightening men" — is "the noblest portion and gift within human reach," wrote the mathematician and philosopher Jean le Rond d'Alembert. He and the philosopher and critic Denis Diderot founded and edited the *Encyclopédie, ou dictionnaire raisonné des sciences, des arts et des métiers* (*Encyclopedia, or a Systematic Dictionary of the Sciences, Arts, and Crafts*); its first volume appeared in 1751, the twenty-eighth and final in 1772, and in the end it consisted of 71,818 articles and 3,129 illustrations. While the *Encyclopédie* was probably the most influential effort of its kind — you can easily trace its radical advocacy for personal autonomy from the rhetoric in its pages, through government efforts to suppress it, to the storming of the Bastille — it wasn't alone. Other encyclopedias appeared in German, Italian, and English (the *Encyclopædia Britannica*).

Not only was this dissemination of knowledge an essential element of the Enlightenment and the Age of Reason, so was the transformation of the audience on the receiving end of the dissemination. A London science lecturer founded the *General Magazine of Arts and Sciences* in January 1755 in order to "bring the world of science within the reach of the 'gentry' of mid-eighteenth-century Britain" — and maybe even "the working classes." By the end of the century, no "maybe" qualifier was necessary. The English *Philosophical Magazine,* first appearing in 1798, declared as its "grand Object" the diffusion of "Philosophical Knowledge among every Class of Society, and to give the Public as early an Account as possible of everything new or curious in the scientific World, both at Home and on the Continent." The founding of the Royal Institution of Great Britain followed a year later; the summary of its mission statement read: "For diffusing the knowledge and facilitating the general introduction of useful mechanical inventions and improvements, and for teaching, by courses of philosophical lectures and experiments, the application of science to the common purposes of life."

The ultimate arbiter of knowledge was no longer an ancient philosopher, a remote deity, a distant ruler. It was you — you in possession of the facts. Transferring new information between the terms of the equation of the universe — reducing x, the unknown and possibly unknowable, by increasing *what-we-know-about-the-universe* — was an exercise in democracy.

Yes, Laplace wrote, Newton's laws "rendered important services to navigation and astronomy; but their great benefit," he emphasized, was to society: the destruction of "errors springing from the ignorance of our true relation with nature; errors so much the more fatal, as social order can only rest in the basis of these relations. *Truth, Justice;* these are its immutable laws." "My own mind," Paine wrote, "is my own church," the principle that Jefferson enshrined as "a wall of separation between church and state" — between superstition and science, the supernatural and the natural, faith and facts.

Look to scripture and you might find reason to believe that Creation occurred on October 23, 4004 B.C., as calculated by the Church of Ireland archbishop James Ussher, or perhaps on the evening of September 12, 3929 B.C., as determined by a contemporary of his, an Old Testament scholar named John Lightfoot (who also pinpointed the arrival of Man: the following day, 9 a.m.). Look to the Earth itself, however, and you might discover layers of sediment that challenge the notion of a planet-submerging flood, or fossils that suggest an older age for the Earth than you would get by counting Old Testament *begats* backward. In his own backyard — literally — no less a literalist than John Hutchinson found fossils that would have been older than the biblical interpretation of the age of the Earth. Either God planted these fossils there for some perverse reason — to drive Hutchinson mad, for instance — or they were the outcome of a natural process occurring over lengths of time previously unimaginable.

Newton had assumed that occasional divine intervention is necessary in order to keep the cause-and-effect machinery of the universe from collapsing, but Laplace's mathematics revealed that through mutually reinforcing gravitational interactions, the planets, moons, comets, and Sun could keep humming along on their own, thank you very much. Laplace, in his 1814 *Philosophical Essay on Probabilities*, took a principle that had been implicit in the concept of universal gravitation from the beginning and made it explicit.

"We ought to regard the present state of the universe as the effect of its antecedent state and as the cause of the state that is to follow," Laplace wrote.

> An intelligence knowing all the forces acting in nature at a given instant, as well as the momentary positions of all things in the universe, would be able to comprehend in one single formula the motions

of the largest bodies as well as the lightest atoms in
the world, provided that its intellect were sufficiently
powerful to subject all data to analysis; to it nothing would be uncertain, the future as well as the past
would be present to its eyes.

The view available to this intelligence is hypothetical. It certainly doesn't belong to the human mind, which "offers, in the perfection which it has been able to give to astronomy, a feeble idea of
this intelligence. Its discoveries in mechanics and geometry, added
to that of universal gravity, have enabled it to comprehend in the
same analytical expressions the past and future states of the system
of the world." But the human mind "will always remain infinitely removed" from that intelligence. The law of gravitation had whittled
the x in the equation of the universe down to this vast intelligence
— First Cause.

The relationship between space and time became even clearer
once astronomers made the connection between a finite speed of light
and distances to celestial objects. "I have looked further into space
than ever human being did before me," William Herschel, the discoverer of Uranus, wrote in 1813. "I have observed stars of which the light,
it can be proved, must take two million years to reach the earth."

The principle easily extended to other sciences. The Scottish
geologist Charles Lyell made this connection explicit, in the early
1830s, in the three volumes of *Principles of Geology*, the *Principia* of
its own field. Lyell put the present's relationship to the past right in
the subtitle: AN ATTEMPT TO EXPLAIN THE FORMER CHANGES OF THE
EARTH'S SURFACE, BY REFERENCE TO CAUSES NOW IN OPERATION.

"In vain," Lyell wrote in the Concluding Remarks, "do we aspire to assign limits to the works of creation in *space*, whether we examine the starry heavens or that world of minute animalcules which is revealed to us by the microscope. We are prepared, therefore, to find that in *time* also, the confines of the universe lie beyond the reach of mortal ken." He was even willing to extrapolate from this insight into another field: "The disposition of the seas, continents, and islands, and the climates have varied; so it appears that the species have been changed."

It certainly does appear that way, wrote Lyell's good friend Charles Darwin in his 1859 *On the Origin of Species*. "Authors of the highest eminence seem to be fully satisfied with the view that each species has been independently created," Darwin wrote in the final pages. Such thinking, though, is a return to the discredited logic of starting with First Cause. "To my mind it accords better with what we know of the laws impressed on matter by the Creator, that the production and extinction of the past and present inhabitants of the world should have been due to secondary causes, like those determining the birth and death of the individual." (Secondary causes and only secondary causes, he might have added — as he did in a letter to a friend: "I would give nothing for the theory of Natural Selection if it requires miraculous additions at any one stage of descent.")

At this point in his text, Darwin invoked the inspiration behind all his investigations. The final words of his book would be "endless forms most beautiful and most wonderful have been, and are being, evolved," but earlier in the sentence Darwin prepared his readers by invoking a comparison they would know well: Species not only have done, and are doing, their evolving, but they've done it and are doing it "whilst this planet has gone cycling on according to the fixed law of gravity."

The fixed law of universal gravitation: one law that unites all. Thanks to gravity, the New Philosophy — natural philosophy — had been a success. Gravity had given the New Philosophy its method, and it had served as natural philosophy's model. It had inspired the solution to countless mysteries. In physics the search had narrowed to "one or another nook," according to one professor. In astronomy, said the Canadian-American astronomer Simon Newcomb, "We are probably nearing the limit of all we can know." The evidence throughout the sciences, read the lead article in the October 20, 1860, edition of *Scientific American,* is "that all the forces in nature are the same thing; merely *matter* in *motion.*" Thanks to gravity, science had chased the *x* in the equation of the universe to near extinction.

And yet —

The German philosopher and physicist Ernst Mach was not the voice of Reason. By the time he had risen to prominence, in the 1870s, the Enlightenment and the Age of Reason were long past.

But Mach was *a* voice of reason — a lone voice at times, arguing for greater rigor in the methodology of science. Yes, that methodology had been successful, and yes, the universal law of gravitation had served as its model. Yet Mach had to ask — and did ask, in an 1872 book — wasn't there a problem here? "The Newtonian theory of gravitation, on its appearance, disturbed almost all investigators of nature because it was founded on an uncommon unintelligibility" — a cause-and-effect relationship across great distances that involved no physical contact. Now, Mach wrote, the theory of gravitation had disguised its philosophical shortcomings by proving its reliability and usefulness. But the philosophical shortcomings hadn't gone anywhere. They'd just gotten respectable. Gravitation, he wrote, "has become *common* unintelligibility."

GRAVITY IN EXCELSIS

On the night of April 17–18, 1955, a one-woman vigil was taking place in a private room at Princeton Hospital. Shortly after 1 a.m., the patient began to mumble; the nurse leaned closer. But the old man was speaking German, which she didn't understand, and so the meaning was lost on her. The patient took two deep breaths; then he was gone.

Albert Einstein, of course, wasn't going anywhere. His legacy, like Newton's, would outlive him; it will endure for as long as physics matters. Einstein was the face of physics — the face of science: not just a fright-wigged iconoclast but someone whose intellect alone could reimagine the universe. For decades the news media had deified him, first with reports that, through his two theories of relativity, he had forever changed our understanding of the cosmos, then with reports, inaccurate yet emblematic of his stature in the popular imagination, that he had developed another grand theory, a sequel of sorts to his greatest hits.

Einstein had been intimately familiar with Ernst Mach's writing. In the early years of the century, shortly after graduating from Eidgenössische Technische Hochschule in Zurich, Einstein participated in a book group that read the day's leading philosopher-scientists (as the New-Philosophers-turned-natural-philosophers were now calling themselves); Mach emerged as the one who most impressed Einstein. (Years later, Einstein would make a pilgrimage to a Vienna suburb to pay his respects to the elderly and ill philosopher.) When Mach made the distinction between uncommon unintelligibility and common unintelligibility, he was suggesting a contrast between new concepts that make no sense within the current understanding of nature, so we challenge them, and the same concepts once they've proven themselves to be reliable predictors of reality: They still make no sense, but we accept them.

"In glancing over the history of an idea with which we have become perfectly familiar," Mach once said, "we are no longer able to appreciate the full significance of its growth." The example he used involved seventeenth-century conceptions of light and gravity. Newton thought light propagated like pellets, or what he called corpuscles. Huygens argued that interpreting light as waves would explain how it spreads out even after it passes through a small aperture —a conception, Mach wrote, that "was incomprehensible to Newton." Likewise, Newton's idea of action at a distance was "unintelligible to Huygens." But, Mach continued, "a century afterwards both notions were reconcilable, even in ordinary minds."

Now the pattern was repeating itself, though Mach wouldn't live long enough to see it to fruition. He died in February 1916, only three months after Einstein introduced the general theory of relativity—a conceptual achievement that was about to make gravity uncommonly unintelligible once again.

The story goes that one day in 1907 Einstein was standing at his clerk's podium in the patent office in Bern, staring out the window, daydreaming about a worker falling from the roof of a house. Einstein stopped him there, and he held him there: a man in midair.

As with Newton's apple in the orchard, the details of this modern myth are open to question and interpretation. Nonetheless, Einstein did experience a sudden insight. He was a full-time patent clerk but also a part-time physics theorist; two years earlier he'd published four formative papers that were just coming to the attention of the deepest thinkers in Europe. As a theorist Einstein knew that in constructing a thought experiment he could imagine ideal conditions. Just as Galileo had exiled air resistance when considering falling weights, or burrs when conjuring frictionless wooden chutes, so Einstein situated his roofer in a vacuum. The man in midair, feet neither on the roof nor on the ground, would have no way to know whether he was falling toward the Earth or whether the Earth was rising toward him.

For Einstein, this kind of equivalence was the physics version of ambrosia: two seemingly unrelated events actually being the same. One of those 1905 papers, "On the Electrodynamics of Moving Bodies," opened with the observation that the current interpretation of electrodynamics didn't make sense to him. Forty years earlier, the Scottish physicist and mathematician James Clerk Maxwell had formulated a series of equations suggesting that while electricity and magnetism could each act alone, they could also combine into electromagnetic waves — Huygens's waves of "uncommon unintelligibility," which our eyes perceive as light. Yet, Einstein said, the current assumption was that the charge a battery exhibits when in motion relative to a magnet at rest, and the charge a magnet exhibits when in motion relative to a battery at rest, are two separate phenomena. And indeed, if a battery is in motion relative to a magnet, the battery will exhibit an electrical charge, while if a magnet is in motion relative to a battery, the magnet will exhibit an electrical charge. But Einstein argued that the effects are the same — that electromagnetic waves are what the contact between battery and magnet is creating, regardless of which object is in motion and which at rest. Just as a passenger on Galileo's imaginary ship or a bystander on Galileo's imaginary dock can think of the ship traveling downriver or the dock traveling upriver with equal validity, so the battery and the magnet have equal claim on being at rest or in motion.

Einstein then reasoned:

If electricity and magnetism "possess no properties corresponding to the idea of absolute rest," and if electricity and magnetism together create light, then wouldn't light also possess no properties corresponding to the idea of absolute rest?

And in that case, both the passenger on the ship and the by-stander on the shore would have equal claim on the properties of light. A beam of light rising vertically from the deck of the ship would seem to be rising at an angle to an observer on the dock, no matter if the ship is traveling downriver or the dock is travel-ing upriver. And vice versa: A beam of light rising vertically from the dock would seem to be rising at an angle to an observer on the ship.

So far, so Galilean. Galileo hadn't used light in his own thought experiments, but if he had, the principle would have been the same as in Einstein's thought experiment: Observers moving in uniform motion relative to each other can make differing claims on the ap-pearance of reality. Einstein, though, possessed two key pieces of knowledge that made a relativity-related thought experiment rely-ing on light different from Galileo's.

First, the velocity of light is finite. In 1676, more than thirty years after Galileo's death, the Danish astronomer Ole Rømer dem-onstrated that light doesn't emerge up there and immediately reach our eyes down here. By observing the eclipses of Jupiter's moons—as they pass behind the planet from our perspective, disappearing from view—Rømer determined that the farther from us that Jupiter is in its orbit, the more time the "news" of the eclipses takes to reach us; the nearer, the less. The velocity of light may be incomprehensi-bly fast—"It is so exceeding swift," the English astronomer Robert Hooke wrote, "why it may not be as well instantaneous I know no reason"—but it is, in fact, finite.

The velocity of light is finite in Maxwell's equations, too, but it's also, crucially, something else: constant. If Maxwell was right, the speed of light in a vacuum never varies. What varies is the distance between the waves. They bunch up or stretch out, depending on whether you are moving toward or away from the source of the light (or it's moving toward or away from you). Much as Newton had accepted Kepler's guess that the shape of orbits is elliptical, so Einstein considered the constancy of the speed of light that was implicit in Maxwell's equations and figured *Why not?* Then he raised this implicit suggestion to the level of postulate.

The rest was just math, and most of the math was simple — high school–level algebra at best. We deal with velocity all day long, whenever we consider the movement of matter: How fast is something going? Sometimes that something is us: If you're walking, you might be traveling four miles per hour. Sometimes that something is a train: maybe eight hundred kilometers every ten hours. Sometimes that something is a slug moving six inches per minute. And sometimes — at least if you're Einstein — that something is light: 299,792,458 meters, or 186,282 miles, per second.

Whatever the velocity of a particular object might be, the formula is always the same. It's distance divided by time: miles per hour, kilometers per ten hours, inches per minute, and so on. What makes light's velocity different, though, is that it's not just 186,282 miles per second, it's *always* 186,282 miles per second. A person walking four miles per hour can speed up or slow down; a slug moving six inches per minute can pick up the pace or not. But not light. The length of the waves, and therefore the number of waves that cross our line of vision in a given period of time, might change (a change that would affect what color we see). But the velocity of the waves remains the same.

Which was why Einstein's introduction of light into Galileo's thought experiment about the dock moving relative to the ship or the ship moving relative to the dock changed the experiment so radically.

The observer on the ship and the observer on the shore are seeing the same beam of light, but to one observer the path of the light is vertical, and to the other observer it's on an angle. To one observer the beam of light is traveling one distance; to the other observer, a slightly greater distance. If the distance an observer sees the vertical beam of light traveling is one mile, then the time that passes for that observer is 1/186,282 of a second. If the distance the other observer sees the same beam of light traveling on an angle is two miles, then the time that passes for that second observer is 2/186,282 of a second. The two observers are seeing the same beam of light, but because they're seeing it traveling different distances, they're seeing it traveling for different periods of time.

Who is right? They both are. The observer on the proverbial ship can claim that one second has passed while the observer on the proverbial shore can claim that two seconds have passed, and each would have the same claim on reality.

You might argue that time isn't "really" slowing down or speeding up, but that argument just raises the question of what you mean by "really." After a moment's consideration, you might answer that what you mean by "really" is what the time or length would be if you were right up against the object. But that's the point: If you moved right up against the object, you would be entering the object's frame of reference; you would be moving along with it. Just as if you were a conquistador in the Caribbean or Kepler on the Moon, you will have traveled to a different shore. Now turn around. What do you see? The shore you have abandoned — your old frame of reference — and it is now the one in motion relative to you. And if on that shore there happens to be standing someone beaming a light signal — well, you can do the rest.

These discrepancies aren't mere semantic tricks. All we know of the universe is what we know right here, right now — *here* at this point in space, *now* at this moment in time. And all we know right here and right now is the information that reaches us. And that information reaches us through light.

Einstein's preference for the name of this idea was *invar_ententheorie,* because that name emphasized its core principle: "The laws of nature are the same in all inertial systems" — the systems that Galileo identified and Newton borrowed, systems that are either at rest or moving in a straight line at a constant speed. Math makes every person's perception of the universe consistent within itself, even though that perception might disagree with someone else's. What caught the fancy of Einstein's fellow physicists as they began to disseminate this idea, though, wasn't the consistency but the disagreement: While the universe would make sense to everyone in exactly the same way, each experience would be different *relative* to any other.

So a "theory of relativity" it was, but it was also, more precisely, a *special* theory of relativity because it held only in a special — in the sense of specific — circumstance: observers moving in uniform motion relative to clocks and rulers (and vice versa).

What Einstein wanted next was a *general* theory, one that would apply to objects moving in *non*-uniform motion relative to one another. One that would apply, for instance, to a worker falling from a roof.

The falling worker, from his own point of view, can think of himself as standing still. He is, in Newtonian terms, experiencing inertia. From an outside point of view — the point of view of a patent clerk, perhaps, standing at a podium and watching through a window — the worker is falling. He is, in Newtonian terms, experiencing gravitation. And since the roofer's point of view and the patent clerk's point of view are equally valid, and since the roofer and the patent clerk are witnessing the same phenomenon, and since the roofer is interpreting what he's experiencing as the effects of inertia and the patent clerk is interpreting what the roofer is experiencing as the effects of gravitation —

"The happiest thought of my life," Einstein would call this insight — this recognition of equivalence.

Even in Newton's day, natural philosophers studying Newton's equations had noted with some queasiness an uncanny coincidence. In one equation the inertial mass is measuring an object's resistance to change in its state of motion. In another equation the gravitational mass is measuring an object's susceptibility to change in its state of motion. In both equations, the values of the mass are the same. Why? Why would the measure of an object's *resistance* to change — the resistance it has to overcome before it can move forward in a straight line — be the same as the measure of its *susceptibility* to change — the susceptibility to move in a downward direction? Now Einstein knew why: The inertial mass and the gravitational mass are measuring the same thing.

As with electricity and magnetism, and as with space and time, you can consider inertia and gravity (or centripetal motion) two separate phenomena. But you can also consider them to be two versions of a single phenomenon. In uniting the motions of matter *up there* and *down here,* Newton had conceived of a combination of two motions: Galileo's straight-ahead inertial and his own downward centripetal. But straight-ahead relative to what? Downward relative to what?

"Two eruptions of Mount Vesuvius occur at different times but at the same place (that is, at the crater of Mount Vesuvius)," Einstein later wrote in notes to himself, exploring this concept. But "the earth rotates about its axis, moves around the sun, and furthermore, moves together with the sun toward the constellation of Hercules. Therefore, one cannot seriously claim that the two eruptions of Mount Vesuvius occurred at the same location in the universe." Instead, "we can only say: the two eruptions of Mount Vesuvius occur at the same place *relative to the earth* [emphasis Einstein's]." References to locations in time and space make no sense unless they specify those locations in relation to something else. For Newton, the frame of reference was the same throughout the universe: an absolute space — "a giant vessel without walls," in Einstein's words — and an absolute time — "an eternally uniformly occurring tick-tock" audible only to "ghosts everywhere."

Take away the naïve notions of straight-ahead inertial and downward centripetal, Einstein realized, and you can reconceive the two motions as one.

To illustrate the point for himself in a less abstract manner, Einstein removed the roofer from his midair position and placed him inside an elevator — a sort of modern equivalent of the below-decks portion of Galileo's ship: a "laboratory" in which the outside world is unknowable (under ideal conditions, as always). Place the elevator on the surface of the Earth, and we on the outside can say that the worker on the inside is experiencing a gravitational acceleration of approximately 32 feet per second every second. Now place the elevator in empty space and accelerate it at the same rate, and we on the outside can say that the worker is experiencing an inertial thrust of approximately 32 feet per second every second. To the roofer inside the elevator, though, the experiences will be identical: What we on the outside perceive as inertia or gravitation will leave the roofer feeling as if he's standing still.

Even a combination of the two in the right proportion will have no noticeable effect on the roofer. Remove the elevator from the surface of the Earth by a certain distance, thereby lessening the gravitational effect (from our outside perspective), and advance the elevator at the right rate, thereby increasing the inertial effect (from our perspective), and the roofer still won't notice anything amiss. Place that elevator in orbit around a planet or the Sun, and it will appear to be moving (from our perspective) in a line that curves, and that curving line will be describing (from our perspective) an ellipse.

Curved lines, however, were outside Einstein's area of expertise. Lines are, by definition, straight, but Einstein knew that the definition held only in Euclid's geometry — one in which lines are straight and parallel lines never meet. Euclid's geometry arose from an intuitive version of the universe, but it also arose, as Einstein was beginning to realize, from a naïve version of the universe. As a student, Einstein had given himself permission to skip the lessons in non-Euclidean geometry; he considered the subject to be an intellectual indulgence that has nothing to do with the universe we live in. Now, though, he had to concede: Non-Euclidean geometry has *everything* to do with the universe we live in.

The math proved to be even more confounding than he had anticipated. Einstein recruited a longtime friend dating to their university days, Marcel Grossmann, by now a prominent mathematician in his own right but also — more important for Einstein's purposes — one who had bothered to show up for the lectures on non-Euclidean geometry. Yet even after the intercession of Grossmann, as Einstein complained in a letter to a friend, the mathematical subtleties made his struggles with the earlier theory of relativity look like "child's play."

Newton's math was sound, as centuries of experiments had affirmed. If you want to describe the motions of matter, whether *up there* or *down here*, Newton's math worked just fine in nearly all circumstances. But just as the absence of an equivalence between distance and time turned out to be a limitation of Galilean relativity, so the absence of an equivalence between inertia and gravitation turned out to be a limitation of Newtonian relativity. Newton's math couldn't account for *all* circumstances. Other circumstances were out there — at least if, like Einstein, you knew where to look.

In writing the *Principia*, Newton had reached into the past to retrieve the peculiar behavior of Jean Richer's pendulums near the equator to check his math. Within weeks of his vision of a falling man, Einstein realized he could perform the same kind of retroactive validation of his own math with a so-far-inexplicable anomaly in the historical record: the orbit of Mercury. Back in the 1850s,

Urbain Le Verrier, the mathematician who had predicted the position of the new planet Neptune, discovered a discrepancy between Newtonian mathematics and Mercury's motions. Le Verrier had known that capturing Mercury's motions in math would be difficult; Mercury is both the runt among the planets and the one nearest the Sun, factors that place it at the extreme gravitational mercy of the most massive object in the solar system. After twenty years of trying to reconcile his math to the motions, Le Verrier conceded defeat: When Mercury reached perihelion, its closest approach to the Sun, it wasn't precisely where Newton said it should be. It was near to the target—very near—but it was not quite precisely *there.*

Precision is what Newton's theory made possible. Precision is what Newton's theory demanded. But the precision in Newton's theory wasn't quite precise enough, as Einstein had begun to suspect, because Newton's theory wasn't comprehensive.

So his would have to be. Yet even when Einstein arranged to present four lectures on his developing theory before the Prussian Academy in Berlin, in November 1915, he still hadn't wrangled a mathematical explanation for Mercury's motions. Nonetheless, he had a lot to report, even if only in provisional form, and so, for a couple of hours each week, Einstein delivered his latest results.

During the other 166 hours, however, he wondered what he was doing wrong.

Einstein opened his first lecture with the confession that he had "completely lost confidence" in his theory. He returned the next week to continue to document his noble failure. Only then, midway through the month and halfway through his series of lectures, did he make the crucial adjustment to his math—an adjustment that, when he cross-checked it against Mercury's motions, produced a perfect match.

This after-the-fact solution to a lingering problem gave Einstein heart palpitations that lasted for days. It also inspired his confidence in the theory. But as Newton had learned two hundred years earlier, you can't just declare *It is enough* and expect everyone to agree. Einstein understood that if his math was going to generate a consensus among the physics community, it would need to lead to more than a postdictive explanation for Mercury's orbit. It would have to make a prediction that an observation would corroborate (or not).

Fortunately, Einstein had already devised one—one that he didn't think was necessary, so confident was he in his math. But he knew the rules: Make a prediction; conduct an experiment; proceed from there.

Einstein returned to the Prussian Academy the following week and delivered his next lecture in the series. Then he returned for the fourth week to deliver his final lecture. Then he published his theory.

Then, he waited.

The announcement on November 6, 1919, provided one of those distinctly before-and-after moments in the history of philosophy and science—the history of anything, really. The world entered the day as gray as a doughboy and exited it dancing like a flapper. Einstein may have expected the result, but he couldn't have hoped for a better response—one befitting the final formulation of the answer to one of the world's oldest riddles, the motions of matter.

The same service that Halley had performed for Newton, the British astronomer Sir Arthur Eddington had now performed for Einstein. Even while the Great War was under way, Eddington was organizing an experiment that would test Einstein's prediction that the presence of mass creates a curvature in space that we perceive as gravity—an effect we can test during a total solar eclipse. Determine the precise part of the sky that the Sun will occupy during the eclipse, take a photo of the stars in that part of the sky when the Sun isn't there, then take another photo of the stars when the Sun *is* there but its usually overwhelming light is blocked by the Moon. Compare the two photos, and the positions of the stars against the sky should differ, because the paths their light is following along the curvature of space will be different. In the first image, the one without the Sun, the light from the stars will be traveling straight to our eyes. In the second image, the one with the Sun, the light from the stars will be bending around the Sun. Eddington dispatched two expeditions—one to the Brazilian city of Sobral, the other to the South Atlantic island of Príncipe—to photograph the total solar eclipse of May 29, 1919. On November 6, Sir Frank Watson Dyson, who had co-led the Príncipe expedition with Eddington, announced to a special meeting of the Royal Society in London that Einstein's math, not Newton's, had correctly predicted the eclipse results.

The general excitement regarding Halley's prediction of a comet's return had preceded the event; Newton's gravitation to some extent had already "proved" itself, but everyone wanted to make sure.

The excitement regarding Eddington's experiment, in contrast, followed the event. In that respect, the more precise historical precedent might be the pandemonium that greeted the publication of Galileo's *Starry Messenger*. In 1610, a crowd gathered in a public square in Florence to listen to a reading of the book in its entirety. In 1919, telegraphs and newspapers could spread the news faster but with no less emotional immediacy; headlines immediately entered the pantheon of journalistic collar-grabbing, from *The Times* of London's rather prosaic if arresting NEWTON V. EINSTEIN to the *New York Times*'s more poetic LIGHTS ALL ASKEW IN THE HEAVENS.

Part of the theory's charm was its audacity. NEWTON V. EINSTEIN was correct only in the narrowest terms; Einstein had amended, not defeated, Newton. Still, the headline captured the sense that our understanding of the universe had entered a new, jazzier age, one in which you yourself didn't even have to share that understanding in order to join the absurdist fun.

Part of the theory's charm was its opacity. Among scientists, an anecdote quickly circulated. Physicist to Eddington: "You must be one of three persons in the world who understand general relativity." Eddington: [doesn't answer]. Physicist: "Don't be modest, Eddington." Eddington: "On the contrary, I am trying to think who the third person is." And if physicists didn't understand Einstein, what hope did we civilians have? A Rea Irvin cartoon in *The New Yorker* depicted the characters in a typical New York street scene — a laborer, a beat cop, a doorman, a woman in fur, even a dray horse — frozen in various states of deep thought; the caption: "*People slowly accustomed themselves to the idea that the physical states of space itself were the final physical reality.* — Professor Albert Einstein."

And part of the theory's charm was its discoverer. Einstein clearly took delight in his sudden celebrity. Even if he wasn't actually winking, he always seemed to be, as if we were co-conspirators in his good fortune — a foolproof strategy if you want everyone to assume they know what you're thinking without really knowing what you're thinking. Another anecdote: Charlie Chaplin and Albert Einstein, at a Hollywood premiere together, are greeted by roars from the bleachers; Chaplin says, "They cheer me because they all understand me, and they cheer you because no one understands you." Clever, but not quite complete: Whatever anyone might have understood of either of them, everyone *thought* they understood Chaplin and Einstein. Einstein's black chalkboard; Chaplin's white theater screen: two *tabulae rasae*.

The theory's usefulness, however, was not part of its charm. For the public, Einstein remained an icon of the heights to which the human mind might reach. Physicists felt the same way, but they also understood that general relativity held no practical application. It accounted for the behavior of a couple of celestial phenomena under extreme circumstances, and it allowed physicists to think about time and space in an even more unconventional way than Einstein's special theory of relativity, but otherwise?

Aristotle's universe lasted two thousand years. Newton's universe lasted two hundred years. Einstein's lasted less than a decade.

The rise of quantum mechanics helped physicists set aside any lingering interest in general relativity. In the mid-1920s, the German physicist Werner Heisenberg and the Danish physicist Niels Bohr introduced quantum uncertainty into the equation of the universe. To some extent they were expanding upon the kind of equivalence that Einstein favored — in this case, one that he himself had championed in a 1905 paper (the one for which he received the 1921 Nobel Prize in Physics). Einstein posited that while light comes in waves, it also comes in quanta, or discrete units — a concept he borrowed from Max Planck, who had introduced the idea of a quantum of energy five years earlier. Heisenberg and Bohr seized on the idea and found that at the quantum level an observer cannot determine position and velocity at the same time. More worrisome for Einstein, though, was the idea that through quantum processes that nobody understood, two particles could "communicate" across vast distances instantaneously, which is to say, faster than light.

When Einstein called quantum entanglement "spooky action at a distance," he was explicitly invoking the interpretation of gravitational motions that Newton had called a "great absurdity" — action at a distance. But he was also implicitly recognizing that from a philosophical point of view, physics might seem to be going backward. Newton had made his peace with the idea of two bodies interacting without apparent physical contact by assuming that *some* kind of contact must nonetheless be present. And now Einstein had shown that, in fact, there is.

In Newton's physics, matter is active; it moves. In Einstein's physics, matter is also active. In Newton's physics, though, space is passive, a vessel for a mysterious exchange between masses. In Einstein's physics, the mysteriousness vanishes, because space is active: It collaborates with matter to produce what we perceive as gravity's effects. That collaboration removes at least some of the mystery: Matter bends space; space guides matter. The "feeling" of "attraction" isn't just mutual; it's continuous in time and space. You can

trace its effects: An object of matter bends space *here,* and that curvature in space extends to that object of matter *there,* at which point
the motion of *that* matter adjusts.

But even that description doesn't capture the nature of general relativity in its entirety — which is to say, *as* an entirety. Newton thought that time and space were absolute — a Cartesian grid
against which to measure the motions of matter. Einstein treated
space in a more fluid manner — almost literally so. In Einstein's interpretation, the universe behaves like an ocean, seamless as a whole
and ceaseless as a process. You might identify a wave here or a whirl-
pool there, but such phenomena don't exist in isolation. They arise
from their environment, and their environment arises from them.
Rather than pinning a geometer's Euclidean compass to the ocean
floor, as William Blake had depicted Newton doing, Einstein was
watching the currents, eyes wide.

But — crucially, in the emerging interpretation of physics — he
would also be creating currents himself. At the quantum level, any
observation of a subatomic particle will necessarily involve an interaction between the observing instrument and the object under
investigation. Bohr, one of the founders of the quantum interpretation of nature, said that on the largest scales we could ignore such
interactions; one example he cited was a photon slamming into the
Moon. In the same way, Einstein could swim off the coast of New
Jersey, where he relocated in the 1930s, without his motions having a
noticeable effect on, say, the tide level in the Mediterranean Sea. At
the quantum level, physicists could choose to ignore the photon on
the Moon; at the general relativistic level, a whitecap off the Amalfi
Coast. And given the absence of extreme conditions in the universe

to which to apply general relativity, the theory came to occupy a scientific backwater, a languid branch of physics that barely registered in mainstream research.

A primordial eruption out of which the entire universe emerges: How extreme a condition is that?

The idea was right there, in plain sight. It had been there all along, once Einstein published the equations for his general theory of relativity. It moved to the forefront in his *Cosmological Considerations* of 1917, a year later. The problem of Newton's needles — an infinite number of them standing upright for eternity as a symbol of perfect gravitational equilibrium — had never gone away, and now it reappeared in Einstein's math. Newton couldn't miss the logical problem of a universe full of matter attracting matter through gravitation yet not collapsing on itself, and neither could Einstein.

Not that Einstein invoked God, as Newton had. When speaking of metaphysical matters Einstein was prone to referring to a metaphorical deity, a kind of Benevolent Logician, but in physics he tended to invoke God when explaining how the universe *doesn't* work; for instance, he countered the probabilistic nature of quantum mechanics with the quip "God does not play dice with the universe." As for how the universe does work, Einstein left God out of the equation — both the equation of the universe and his own equation for the general theory.

One side of Einstein's equation didn't equal the other, suggesting that the universe was out of balance, either expanding or collapsing. Clearly, though, it was doing neither. So Einstein opted to balance the two sides of his equation by plugging in an *x* of his own, a variable he designated with the Greek letter lambda. It would represent in math whatever was keeping all those needles upright and, on the largest scale, motionless. Einstein didn't know what it was; all he knew was that he needed something to explain the universe's stability. Just as Newton set aside the search for a second cause leading to gravitational effects, Einstein left the puzzle of lambda for future generations to solve.

Unlike Newton's needles, though, Einstein's puzzle *was* solved, and swiftly. First, in 1925, the American astronomer Edwin Hubble announced the discovery that our Milky Way galaxy, a vast collection of stars, was not the farthest extent of the universe; at least a few other galaxies were out there. Two years later the Belgian physicist and priest Georges Lemaître compared two sets of numbers. One set contained the astronomical measures of the distances to the galaxies in his data sample. The other set contained their redshifts — the amount their light has been stretched toward the red end of the electromagnetic spectrum by their motion away from Earth (or, if you prefer, Earth's motion away from them). When he plotted the values of the two sets on a graph, a pattern emerged at once: the farther a galaxy, the greater its redshift. In 1929 Hubble independently reached the same conclusion: The universe is somehow expanding.

The immediate question was inescapable: Expanding from what? Reverse the outward expansion of the universe and you eventually wind up at a starting point, a birth event of some sort. Almost immediately a few theorists suggested a kind of explosion of space and time, a phenomenon that later acquired the name Big Bang.

Newton hadn't needed God, and Einstein hadn't needed lambda, to keep the universe stable after all — because the universe wasn't stable.

Other theorists thought they found flaws in the interpretation of the evidence that the universe is expanding. Rather than being volatile and finite, they suggested, the universe might be stable and infinite, and rather than mysteriously emerging out of a birth event, it might be, no less mysteriously and no less illogically, producing matter. Not a lot. Just a particle here and a particle there every once in a while — and in a beginningless universe not unlike the kind the ancients believed we inhabit, "every once in a while" could well add up to galaxies. This interpretation they called the Steady State.

In July 1965, just ten years after Einstein's death, two articles appeared adjacent to each other in *Astrophysical Journal Letters*. One carried a mathematical prediction. If the universe began in some sort of birth event, four physicists from Princeton argued, then the relic radiation would still be observable, though not in the visible region of the electromagnetic spectrum. Instead, the expansion of the universe would have stretched the light to a specific frequency all the way in the long-wavelength microwave region. In the other paper two astronomers from Bell Labs, also in New Jersey, described an observation of that specific frequency everywhere they could point their radio telescope. For the majority of astronomers and physicists, those two papers decided the dispute in favor of the Big Bang. Their simultaneous publication provided a classic one-two punch of scientific validation: mathematical prediction, observational evidence.

But those papers also provided a clear demonstration of the value of using a new kind of astronomy to investigate the effects of gravity. The optical part of the electromagnetic spectrum — the rainbow that's available to our eyes — is only a narrow sliver of a much wider bandwidth. The idea of using a non-optical part of the spectrum — radio waves, for instance — to perform astronomy was less than two decades old. When Galileo first pointed a telescope at the night sky, he had no reason to think that the heavens would contain more matter than what we had been observing and recording for thousands of years. Yet there it was, and there it had continued to be, with every new improvement to the technology: new moons, new planets, new stars, new galaxies. And now astronomers had begun to point non-optical telescopes at the sky, and while they had no reason to think that their observations would reveal new secrets, let alone insights into general relativity, they had the history of the telescope to inspire them, and they could hope. And the heavens were rewarding their hopes on a regular basis.

Non-optical wavelengths proved to be especially fertile for the investigation of the universe operating under the influence of general relativity. Einstein had anticipated a few experiments that would test the extreme conditions that his mathematics predicted, but he was thinking in terms of optical wavelengths. The universe in invisible wavelengths revealed even more extreme conditions — conditions that fall outside the visible part of the spectrum specifically *because* they are so extreme.

The parts of the spectrum with waves longer than our eyes can see — infrared, microwave, radio — can reveal powerful phenomena from the distant, and therefore earlier, universe. These phenomena would have given off so much energy that, even though they're so far away from us, we can still "see" them, albeit faintly.

Then again, the parts of the spectrum with waves shorter than our eyes can see — ultraviolet, x-ray, gamma — can reveal ultra-powerful phenomena from the nearby, relatively recent universe that haven't yet stretched to any significant degree. But high-energy signals from the distant universe? Those phenomena are the kind that emit such high degrees of energy that, even though their light has traveled from the farthest and oldest reaches of the universe, the waves still haven't dissipated. And those phenomena are the kind that would match physicists' predictions of how matter in the throes of extreme general relativity would behave — predictions that already existed in astronomy's historical record.

Decades earlier, when general relativity and quantum relativity were still somewhat novel, absurd possibilities lurked in the math, and some theorists love nothing more than lurking absurdities. The 1967 discovery of a source that pulsed every 1.3 seconds — a pulsar — matched a mathematical prediction from the 1930s that when stars of a certain mass explode, their cores will gravitationally collapse into a small planet consisting of nothing but neutrons, spinning hundreds of times a second even though each teaspoon would weigh as much as a mountain on Earth. A probe to observe whether the Moon emits x-rays found that it does not; what does emit x-rays, however, is seemingly everything else: In x-ray vision, the optically dark sky is noonday bright. Further observations showed that these objects match another mathematical prediction about collapsing stars of even greater mass than those producing neutron stars: something so gravitationally rich that not even light could escape — a phenomenon that, in 1968, got the moniker that stuck: *black hole*. And not only are black holes common, but one seems to be at the center of every galaxy — and the universe contains at least 125 billion galaxies. And not only is a black hole at the center of every galaxy, but its mass corresponds to the mass of the galaxy, as if one determines the other. For astronomers, as one of the leading researchers in galaxy evolution recently told me, "the two topics are sort of inseparable" — two topics that barely existed a generation ago, and one of which, black holes, nobody really understands.

Talk about common unintelligibility.

❖

Actually, let's.

In the summer of 2014, a series of images began landing in my email inbox. As I opened the files, one after another, day after day, I noticed my shoulders crowding closer to the screen, as if I needed to provide my laptop with winglike protective cover. These files could, I knew, fetch a small fortune.

They were images of the black hole from the soon-to-be-released, 3D-in-IMAX spectacular *Interstellar*. Catnip for the Comic-Con generation.

A Los Angeles–based documentary film company I occasionally worked for had hired me to write a treatment for a short documentary on the making of *Interstellar;* the short would be the kind of five-minute teaser that museums might project on planetarium domes while school groups or rainy-day parent-kid combos try not to nap. (Or, as I can attest as a onetime occasional rainy-day parent myself, try *to*.) The *Interstellar* mini-doc eventually came to nothing, but for a few days I had the privilege of being one of the first people in history to have seen a true image of a black hole — *true* meaning what we know we would see if we could see it in real life, which is to say, what we know from math.

The idea for black holes is a direct result of Newton's law of universal gravitation. In 1784, nearly a century after the publication of the *Principia*, the English clergyman and amateur astronomer John Michell suggested that if light were matter, as Newton thought, it would have to follow Newton's laws. And if Newton's laws were correct, an object large enough to contain sufficient mass could overwhelm light's mass, creating a "dark star." Fifteen years later Laplace provided a mathematical foundation for such an object, but that same year, the polymath Thomas Young demonstrated that light acts as a wave, and Laplace abandoned the idea.

Einstein's 1905 paper on the photoelectric effect, however, suggested that light travels as both waves and packets of matter — photons. Shortly after Einstein formulated the general theory ten years later, the German astrophysicist Karl Schwarzschild found a solution for Einstein's equations that took the math to its logical — or illogical, depending on how much value you place on intelligibility — extreme. An object that collects enough mass inside a small enough radius will have an escape velocity the speed of light. At that point, the object will also collapse into itself, disappearing from view (from the perspective of the outside world) and leaving behind only its gravity. An object doesn't need to be huge to trap light, as Michell and Laplace had assumed; it just needs to be sufficiently dense.

Impossibly dense, thought many physicists, including Einstein. Such an object could result only from mass collapsing into a state of infinite density — what physicists call a singularity. And infinities don't lend themselves to enthusiastic scientific endorsements. Yet there the solution was. It was in the math, and for three centuries, "in the math" had proven to be a reliable predictor of reality.

Another story — and what are these stories of great minds reaching great conclusions if not creation myths, even if they're true? — says that a few weeks before the Royal Society announcement about the 1919 eclipse expeditions, a telegram arrived at Einstein's office at the University of Berlin revealing the result of the experiment. He showed it to a student, and when she congratulated him, he said, "But I knew that the theory is correct." And what, she asked, if the result of the experiment had contradicted his calculations? "Then I would have been sorry for the dear Lord," Einstein answered. "The theory *is* correct."

Einstein's attitude had been the same after the publication of his special theory of relativity. Einstein, like other physicists dating back at least to Newton, had placed his faith in math, and math had rewarded him. In 1905 his math showed him that time did not exist independently of space; the two were co-dependent, rendering the measure of time relative. Within the next couple of years, however, Einstein kept hearing about an experiment that disputed his claims. His response was a shrug: The math is right; the experiment is wrong. And so they were: The flaw belonged not to the math but to the experiment.

Even so, Einstein knew that however right the math might seem, it still might be wrong. In 1914 two eclipse expeditions set out to test an earlier formulation of Einstein's general theory, one that

Einstein was "completely confident" in, as he said in a letter to a friend, yet also one that gave half the value to the crucial measurement of star displacement as the later version of his theory. Fortunately for Einstein — if not the good Lord — one of the eclipse expeditions fell victim to the weather, the other to the border restrictions of the new war.

Human frailty, however, could extend beyond mistakes in math. The imagination — even the imagination of someone who could conjure thought experiments with seeming effortlessness — simply might not be up to the task of reconceiving the universe.

Sometimes the human imagination can't — or at least fails to — conceive of a new phenomenon. When Einstein invoked lambda to explain why the universe was neither expanding nor contracting, he was making the same conceptual error that Aristotle had made in relying on Egyptian and Babylonian observations from one or two thousand years earlier: He wasn't thinking on a big enough scale of either time or space. In 1916, the universe consisted in its entirety of what we today would call our galaxy, and that universe looked, on the whole, stable. So despite what his math was telling him, Einstein made the universe stable. After examining Hubble's data on the expansion of the universe, Einstein offered his famous concession that lambda was his "greatest blunder," and then he moved on — as, by then, had physics.

Even so, Einstein still hadn't learned his lesson. In the 1930s he expanded upon the discovery that the great mass of the Sun reroutes space sufficiently that we can see stars behind it. Would other stars — stars of the kind that once belonged to the farthest aethereal sphere orbiting Earth and numbered maybe six thousand (by the naked eye) but were now simply uncountable pinpoints of light extending as far as every new generation of telescope could see — perform the same favor for astronomers? He wasn't suggesting that current telescope technology was able to detect such an effect, and he doubted that any future technology would allow it. But as was the case with assuming the universe consisted of one galaxy with stable moving parts, Einstein wasn't thinking on a big enough scale. Younger minds, however, had come of age in a different universe — Lemaître and Hubble's many-galaxied expanding universe. The Swiss-American astronomer Fritz Zwicky suggested an alternative to using a star as a lens: using a galaxy. Current telescope technology wouldn't be able to detect that effect, either, but Zwicky assumed that an observation by future generations was no longer out of the question. It came in 1979, and since then gravitational lensing using individual galaxies or clusters of galaxies has become a routine tool for astrophysicists examining the distribution of mass in the universe.

The gravitational mass of a black hole would perform a similar trick of the light, a pyrotechnic display that the director of *Interstellar*, Christopher Nolan, wanted to capture as accurately as possible. To supply the math he turned to Kip Thorne, perhaps the premier black-hole theorist of his day as well as the gravitational-wave physicist who, in one of our conversations over the years, would call my "What is gravity?" question meaningless.

Thorne had been working on the math for decades, using Einstein's equations to understand the behavior of black holes — their spin rates, their temperatures, and so on. Now, though, he was feeding his math into computers not to discover how black holes behave but to find out how they would look on the big screen. And the computers on the receiving end of his math weren't Caltech computers, as powerful as they might be; they were *Hollywood* computers.

Nolan had commissioned the same CGI house that provided effects for another of his films, *Inception*. Double Negative, or DNeg, had worked on more than one hundred movies in the previous fifteen years; its cityscape-folding contributions to *Inception* won the Oscar for Best Achievement in Visual Effects. DNeg took Thorne's data and brought it to digital life — the schematic models, stills from the film, and test footage that were soon flooding my laptop.

Black holes are, in some ways, simple. They consist of two parts. I had instantly recognized the black sphere at the center of the image as one of the parts: the event horizon, the "bubble" of blackness. At its heart resides the second component of a black hole, a singularity. I also recognized the black hole's two most obvious effects on its environment: the accretion disk, a ring that gathers around the "equator" of a spinning black hole as it sweeps up the gas from its galactic neighborhood; and the gravitational lensing of the background stars and galaxies as, from our point of view, their light follows radical curvatures in space-time to reach our eyes as smears and arcs (though, as is true for you as you cross a street, the light "thinks" it's going in a straight line).

Not until I spoke to Thorne, though, did I realize what I wasn't recognizing about the image — what made it different from every other cinematic representation of a black hole as well as the vast majority of previous scientific models.

Think about Saturn. Like a black hole, Saturn has a ring around it. (Rings, actually, but let's simplify here.) From our point of view we see only the portion of the ring that is on our side of the planet. The rest is out of sight on the far side of Saturn. But in the schematic model of *Interstellar*'s black hole, the ring never goes out of sight. The section on the near side of the black hole is visible, of course, but so is the section on the far side. Rather than being out of sight, it rises up over the "top" and curves under the "bottom" of the event horizon.

Of course it does! I thought when I finally realized what I'd been looking at for days. Even before I started the project, I knew that the distortions from a black hole's gravitational lensing would be so great that an observer on one side would be able to see objects located on the opposite side. Then when I did start the project, I recognized the distortions in the background stars as light that's curving around the black hole. The idea of gravitational lensing is so counterintuitive, though, that in my mental inventory of what would be visible even though it was on the other side of the black hole and therefore, in principle, "out of sight," I'd neglected to include the rest of the accretion disk.

That bright arc at the top? That's the accretion disk. That bright arc at the bottom? Same disk. (Gravitational lensing not only bends light but multiplies images.)

Thorne told me he'd experienced his own version of an *Of course!* moment. As a black hole expert, he knew that the out-of-sight portion of the accretion disk should be visible. But when physicists study the behavior of black holes, the far side of the accretion disk is unnecessary data, so they leave it out. He hadn't been thinking about that effect when the files started arriving on his own computer. Now, for the first time, Thorne was seeing that effect in all — literally all — its glory.* †

My surprise didn't surprise me. What did I know? Thorne's surprise did surprise me, but it was, in the end, only the surprise of someone who needed a reminder. What surprised me more, though, was the response I got from an astrophysicist friend when I wrote about this effect online. "Richard," he emailed me, "I never thought of that, but once you pointed it out — of course!" His surprise was the surprise of revelation: of accepting an unintelligibility as common because it's *in the math*, only to realize you've forgotten just how uncommon it is.

But even if the human imagination is able to reconceive the universe by conceiving a new phenomenon, and even if that new phenomenon is in the math, should we believe it *just because* it's in the math?

Around the same time that I was doing the research on *Interstellar*'s black hole, I found myself watching the 1991 documentary *A*

---

* Despite having gone to such lengths to ensure scientific accuracy, Nolan chose not to include that information in the movie, presumably because it isn't essential to the plot: The characters don't need to know what they're seeing. Neither do we in the audience, I suppose. But I did know, and I have to say, it added a layer of wonder to the viewing experience.

† A 1979 paper in *Astronomy and Astrophysics* by a French astrophysicist and poet with the improbable surname of Luminet (see the bibliography) described the effect and even included illustrations, but its impact, whatever it might have been, has faded since then.

Brief History of Time, based on Stephen Hawking's 1988 book. In the film, Hawking invites us to imagine what we would see if we were observing an astronaut nearing a black hole's event horizon. The astronaut is wearing a watch, Hawking says, and the second hand is ticking toward 12:00. As the astronaut gets closer to the event horizon, the motion of the second hand will appear to us, observing from a distance, to be slowing down. The closer the astronaut gets to the event horizon, the slower the motion of the hand on the watch, from our perspective. "Each second on the watch would appear to take longer and longer," Hawking says, "until the last second before midnight would take forever."

So far, so commonly unintelligible. The idea of the image of the person freezing at the lip of the event horizon long ago entered even the popular literature on the subject of general relativity. The image will freeze, and its radiation — the photons that reveal the image to us — will eventually lose energy, entropy being what it is, and fade.

Then the documentary reverses the point of view and explores what the astronaut would be experiencing. That poor sap has a perspective, too. And that's where things get weird. (Well, weirder.)

Just as the outside observer would see the astronaut's time as going slower and slower, so the astronaut would see the time in the rest of the universe as going faster and faster. The closer the astronaut gets to the accretion disk, the faster the passage of time. At the moment the outside observer sees the astronaut's time come to a halt, the astronaut would witness the entire future of the universe.

Of course!

Yes, I know: the entire future of the universe. Still: of course.

Of course because it makes logical sense. The laws of physics that apply to us also apply to the astronaut. And if the math behind the laws of physics says that for us his time will appear to slow down and stop, then for him our time will appear to speed up and — whatever.

That *whatever* can be troubling. When science begins to strain the language, we can be pretty sure we're getting into intellectual concepts that we've never before had to put into words. Even so, somebody at some point came up with the words *time dilation* to describe the different perceptions of time for two observers in different gravitational settings. The idea has become so common that much of the plot of *Interstellar* involved time slowing down or speeding up, depending on the characters' proximity to or distance from the black hole — a concept Nolan didn't even bother to explain. He trusted that the audience would accept time dilation as a matter of scientific fact.

But this particular *whatever* — seeing the entire future of the universe — was new to me. Was I alone? I wrote about it online, said it had blown my mind, and asked readers, "Does it, or does it not, blow your mind?" A response arrived in the Comments section later that day: "Blownless. General relativity is a hundred years old."

So: a common unintelligibility.

And yet . . . the entire future of the universe?

Maybe my human imagination wasn't as strong as that reader's. Or maybe something was wrong in the math. Or maybe some way that we were thinking about the subject was off in a way we couldn't yet recognize. "Everyone's calculations show that the universe started from a singularity," says one theorist specializing in cosmological considerations, "but no one believes it." I once saw a talk in which a physicist needed to characterize the origin moment; using scientific shorthand, he designated it as "$t = 0$" — time equals zero. "Whatever that means," he added.

So: common unintelligibility, with a caveat — an asterisk that, one hundred years after Einstein reconceived our universe in ways even his imagination couldn't fathom, says, *But, just between us,* "*uncommon.*"

GRAVITY IN OUR BONES

The newly knighted PhD in physics kneels before the professor and, awaiting a final benediction, bows. The professor raises one hand over the student's head. The hand is holding an apple.

"One, two, three," the professor says, and he lets go.

The apple does what apples — objects — do. It falls toward the center of the Earth. But before it can get there it hits two obstacles, first the student's head, then the surface of the Earth, in this case a sidewalk.

The ceremony takes place every spring in front of a monument on the campus of Tufts University, just outside Boston. The early twentieth-century economist Roger Babson underwrote the monument, as well as similar monuments on more than a dozen college campuses across the Northeast (including Babson College, which he founded). Each carries an inscription that, Babson hoped, would impress upon future generations the urgent need to perform gravity research — or, more accurately, anti-gravity research.

Gravity, to Babson, wasn't just an enemy. It was *the* enemy. "Gravity — Our Enemy Number One" was the title of an essay he wrote that recounted the source of his crusade:

> When I was a boy my oldest sister was drowned
> while bathing in Annisquam River, Gloucester, Mass.
> Yes, they say she was "drowned," but the fact is that,
> through temporary paralysis or some other cause
> (she was a good swimmer), she was unable to fight
> Gravity, which came up and seized her like a dragon
> and brought her to the bottom. There she smothered
> and died from lack of oxygen.

Babson wrote the essay in 1948, a year after he also lost a grandson through drowning, but by then he had already acquired a reputation as a bit of an eccentric. During the Depression he commissioned out-of-work stonemasons to engrave words of wisdom on boulders in the forests near the northern coast of Massachusetts. Some of the inscriptions were of the sort that might indeed inspire the wayfaring unemployed who somehow found themselves tramping through these particular wilds: IF WORK STOPS VALUES DECAY; KEEP OUT OF DEBT; GET A JOB. Others were merely sentimental evergreens: COURAGE; KINDNESS; HELP MOTHER.

Babson's obsession with Isaac Newton, though, was sui generis. He titled his autobiography *Actions and Reactions* in honor of his own application of Newton's cause-and-effect physics to financial forecasting. He collected one thousand editions of Newton's publications, many bearing Newton's autograph or annotations. Among other Newtonia he amassed were a Newton death mask that once belonged to Thomas Jefferson, the entire fore-parlour of Newton's last London residence, and the descendants of trees from Newton's own apple grove. The review of *Actions and Reactions* in the March 1, 1936, *New York Times* began: "Roger W. Babson, known throughout the world as an expert on finance and financial statistics, emerges in the pages of his autobiography as a man who belongs in the class of American leadership that holds such figures as Edison, Ford and Coolidge." Edison, a longtime friend, once advised Babson that an anti-gravity material might be "coming about from some alloy." Babson dutifully installed three inspectors in the US Patent Office who, he wrote in *Actions and Reactions,* "were constantly on the watch for any machine, alloy, chemical or formula which directly relates to the harnessing of gravity."

The professional skeptic Martin Gardner devoted a chapter of his 1950 book *Fads and Fallacies* to Babson, and especially to the Gravity Research Foundation, which Babson had founded in 1948. The foundation held conferences, administered an annual essay competition,* and sponsored the installation of the monuments on college campuses. Gardner pointed out that Babson's idea for some

---

* Which continues to this day: Five winners receive awards ranging from $500 to $4,000. The focus of the research, however, is no longer anti-gravity but the physics of gravity, and serious researchers take the competition seriously — or at least they take the potential for a payday

sort of "gravity screen" was scientifically problematic because "gravity is not a 'force' which pulls objects to earth, but rather a warping of the space-time continuum." In which case, Gardner continued, a "'screen' between apple and earth would have no effect for the simple reason that there is no force to be screened off."

Carriage return; new paragraph — a typewriter maneuver as clean and swift as an executioner's blade:

"If Babson is aware of all this, he remains blithely undismayed."

Yet Babson also had a valid point. Maybe not the point he was making in the first part of the inscription on the monument at Tufts —

IT IS TO REMIND STUDENTS OF
THE BLESSINGS FORTHCOMING
WHEN A SEMI-INSULATOR IS
DISCOVERED IN ORDER TO HARNESS
GRAVITY AS A FREE POWER

— but this point, which immediately follows:

AND REDUCE AIRPLANE ACCIDENTS

A French chemistry and physics teacher named Jean-François Pilâtre de Rozier might have agreed. On October 15, 1783, he watched as Jacques-Étienne Montgolfier and his brother Joseph-

---

seriously. Stephen Hawking, for instance, won an award six times, including first place in 1971.

Michel became the first humans to fly. They climbed aboard the basket of a balloon they had designed, workers threw aside the ballast, and they ascended from the suburban Paris soil. They rose a few dozen feet, then returned. De Rozier ascended in the same balloon the same day. In both flights, however, ropes tethered the balloon to the Earth. The following month de Rozier earned the distinction of being one of the first two humans to fly untethered, when he and a companion rode a balloon over the length of Paris. Two years later de Rozier earned the further distinction of being one of the first two humans to die in an air crash, when the balloon he and another companion were trying to pilot across the English Channel plummeted back to Earth even before leaving France.

That outcome was hardly a surprise. One doesn't need Babson's boulders to intuit the danger of ballooning over France or even bathing in the Annisquam River. We know it like we know our relationship to the horizon: without having to think about it — though think about it many of us do, as whenever an airplane-and-us inertial frame accelerates down a runway, straining to uncouple itself from the Earth-and-airplane-and-us inertial frame, and we seek comfort in the knowledge that the sky is already lousy with Icaruses.

The PhD pantomime at Tufts is a joke, of course — an exercise in irony. Yet there's something poignant in the exercise. The new physicists have spent years learning about how the general theory of relativity applies to the universe, but the final stage of their ceremonial entrance into the world of professional physicists is a reminder that our primary relationship to gravity is still with Earth. As Cicero's elder Scipio might have scolded them, "How long will your mind be fixed upon the ground? Do you not see what temple you have entered?"

Well, yes and no. Cosmologists might occupy a temple that the rest of us can hardly imagine, but they can hardly imagine it themselves. When they speak of the mathematics of gravitation, they divide the universe into "Big G" and "little g" — *G* being a universal constant* and *g* being the local rate of acceleration. That distinction serves a valid scientific function; *G* represents the cumulative effect that theorist John Archibald Wheeler once described as spacetime telling matter how to move and matter telling space-time how to curve, while *g* represents, loosely speaking, how fast things fall. Still, it *is* a convenience, another instance of *up there/down here*, a reflection of a fundamental psychological divide that might never go away, since they're — we're — only human: We know that Earth orbits the Sun, but at the end of the day, we still say that the Sun goes down, and when we think about gravity, what comes to mind isn't black holes or the Big Bang but airplanes and apples and us.

Which is why the caveat of *But, just between us, "uncommon"* is particularly vexing — perverse, almost cruel. The temple of the universe has gained many chambers of complexity over the course of three or four centuries, and especially over the course of three or four decades; the *universal* part of the universal law of gravitation has blurred the boundaries between *down here* and *up there* — the

---

* In Newtonian terms, *G* correlates masses with the inverse square of distances, and in Einsteinian terms, the geometry of space-time with the energy-momentum tensor.

mythological boundaries, and the cosmological, and the psychological. Yet it has also reinforced those boundaries, even adding several of its own, and so in some ways we find ourselves back where we began: asking where we stand in relation to the universe.

For a start, we still do just that — stand.

Falling over is easy. All you need to do is nothing.

Staying upright is hard. It requires your body to do the opposite of what gravity would have it do if you left nature to its own devices. You can think of "staying upright" as "standing" — which it is — but that formulation is really just a positive iteration of what is essentially a negative action: "not falling over."

As a species accumulates all manner of incremental changes through evolution, those that are anatomical have to account for the effects of gravity, until eventually the relationship of anatomy to gravity is each species' own. A giraffe's pulmonary system needs to pump blood to a greater height than a human's head, so its anatomy requires greater blood pressure and more resilient blood vessels. At the other extreme, an insect interacts with gravity at a minimal level. It can coast on a breeze for great distances without fear. (What it needs to worry about instead is surface tension; a puddle is the arthropod equivalent of the Annisquam River.)

The incremental anatomical changes that *Homo sapiens* has accumulated through evolution have cultivated the body's ability to remain upright for prolonged periods. The visual system registers both external motion and where the head and body are in relation to the world. Sensors in our muscles, tendons, and joints — the proprioceptive system — record the positions of our legs and feet relative to the ground as well as the position of our head relative to our chest and shoulders. Organs in the inner ear — the vestibular system — are sensitive to the position and movements of the head. All this information feeds into the brain stem, which is also gathering knowledge of previous similar experiences from the cerebellum and cerebral cortex. Once the brain stem has integrated past and present data, it coordinates the actions of the various parts of the body.

Now set this matter — a physiological system of nonstop measurement-taking and correction-making as elaborate and complex as Laplace's solar system — in motion.

Snakes, naturally, cheat. Not that they're trying to walk, exactly. What they're trying to do is weave, by alternately contracting and relaxing their muscles in waves, which is also how fish move. But a snake's body hugs the ground, letting gravity do all the dirty work. Quadruped reptiles hug the ground, too, but they also possess legs, and they use them to move. Quadruped mammals don't touch the ground while they walk except, of course, through their legs, but they have the advantage of owning one more leg than the minimum of three that are necessary for a stationary object to independently remain upright.

Human legs, however, number one fewer than is necessary. Tripods are better at remaining upright than bipeds. But then, tripods can't walk. Humans manage not only to remain upright but to do so while locomoting.

I once worked on the script for one of those big-screen National Geographic movies. The title was *Robots 3D,* and the working subtitle (which didn't make the final cut) was *It Isn't Easy Being Human(oid)!* As that summation suggests, the theme of the movie was that designing robots to do what humans do is so complicated because what humans do is so complicated.

Intelligence, I wasn't surprised to hear, is notoriously difficult to replicate. The same, though, is true of walking — which was somewhat surprising to me, not least because I had no idea the difficulty was the same as it is for artificial intelligence: that science doesn't fully understand how we work. If you've ever seen one of those film clips of robots "playing" soccer or playing "soccer" (placing the air quotes is tricky because the robots seem to be doing something in the vicinity of a soccer ball that involves falling over),* you'll know what I mean. The first generation of robot legs generally swung from a hip joint, which is precisely what human legs would do, too, if our anatomy worked like a six-year-old's stick-figure drawing.

Instead, our ability to walk relies on a kind of evolutionary jujitsu: We overcome gravity by surrendering to it. *Do your worst,* is our attitude. And gravity does. But then we're, like, *Sucker!* If you think of what a soccer-playing robot does as an uncontrolled fall, then you can think of what we do as a "controlled fall" — which is in fact a term scientists use for what the rest of us call "walking."

Let's start with the left leg. Suppose it's upright and unbent, and it alone is supporting the body mass. We're standing on that leg, which is to say, we're not falling off that leg. Then we lean forward, and we let go. We give up. We yield to gravity, allowing it to do to us what it does, while we do nothing to stop it.

---

* Grossly unfair characterization. Also, not inaccurate.

We fall.

Now suppose that at the same time we're falling forward, the right leg begins to extend in the same direction. At first it's bent at the knee, but the more the leg stretches, the more the knee unbends. At last the leg has stretched until it's straight, which is the moment that the downward trajectory of the foot reaches the ground. Which is the moment we begin to overcome gravity.

We rise.

Not by a lot. By just enough to return us to the same height as when all the weight rested on the upright, unbent left leg, only now all the weight rests on the upright, unbent right leg. Surveying the landscape from this perch — the height that we list on our driver's license — we funnel data from the visual, proprioceptive, and vestibular systems into the brain stem. And then once more we lean forward, and once more we do nothing. And once more:

We fall.

In the course of normal events, we don't think about rising and falling. We just do it. It's a part of our natural rhythm. It's how we experience the universe on the scale of our species. "Not falling over" is our common stance on our common ground. It transcends what we today would call science; it recalls the realm of the philosopher-scientist, the natural philosopher, the New Philosopher, the old philosopher, and the mythmaker.

The American philosopher Mel Brooks once said, "Tragedy is when I cut my finger. Comedy is when you fall into an open sewer and die." The equally estimable British philosopher Alfred Hitchcock delineated the same distinction, the "fine line between tragedy and comedy," during a 1972 television interview. He invoked the same example that Mel Brooks had: "the old-fashioned scene of the man walking toward the open manhole cover." In his deliberate, deadpan manner, Hitchcock established the stagecraft. The man is wearing a top hat, and he's reading a newspaper, and he's walking — "and suddenly he disappears down the hole! And everybody roars with laughter. But" — Hitchcock leaned forward, as if to peer down the hole — "suppose you took a second look. His head is cut. He's bleeding. You send for an ambulance."

Cut!

Hitchcock straightened his posture and summarized his argument: "Slipping on the banana skin can be very painful."

In their day jobs, Brooks and Hitchcock were filmmakers; that they both chose a visual example to illustrate the difference between comedy and tragedy is no coincidence. But it's also no coincidence that they chose the *same* visual example. Falling down a manhole is a universal symbol for a pratfall — right up there (or down there) with slipping on a banana peel. Not that the fear of falling down a manhole or slipping on a banana peel is universal. But the possession of a prat is, and so is the fear of falling on it.

Hitchcock was right: Drop someone down a manhole and he might survive. If you're a comedian or a tragedian and you want to raise the dramatic stakes, you need to think on a bigger scale, a grander scale. A *higher* scale. So you forsake the street for a cliff, because a cliff is, for your purposes, a manhole writ high. And if the cliff is writ sufficiently high — and the cliff must be sufficiently high, or what's the dramatic point? — the outcome of a character's being on it is stark: life or death.

In comedies, the threat of falling isn't real; in fact, the emptiness of the threat is part of the joke. Cartoon characters are forever falling off cliffs, and they're forever somehow surviving. A coyote-outlined crater in the canyon floor of a Road Runner short doesn't signal the permanent exit of the antagonist; it's merely the transition to the next scene, where Wile E. Coyote will return, intact. In the 1941 Merrie Melodies short *The Heckling Hare*, Bugs Bunny and Willoughby the dour dog fall off a cliff, and they keep falling. Enemies only moments earlier, they now cling to each other, screaming and falling, screaming and falling. For forty-one seconds they fall — possibly the longest fall in movie history. Just before hitting the ground, though, they pull up short, right themselves so that they're descending feet first, and, gently, land standing.

"Nyah, fooled ya, didn't we?" Bugs says to us, breaking the fourth wall. Nyah, not really. If the two of them had died, *that* would have fooled us. The punch line, instead, is *how* they cheat death. They do the impossible: They deny gravity.*

Characters on cliffs in dramas play by different rules. If they are to survive, they have to do so the hard way. Unlike cartoon characters, they can't deny gravity. They have to *defeat* gravity. They can't behave as if a fall doesn't have fatal consequences. They can't land

---

* The same principles apply to human characters, but only if they're sufficiently cartoonish — for example, Pee-wee Herman and an escaped con driving off a cliff in *Pee-wee's Big Adventure* in a sequence so similar to *The Heckling Hare* that it might as well be an homage. And maybe it was.

standing up. They have to *not fall* in the first place. They have to hang on for dear life. And even that's not good enough.

When authors of serial novels in the nineteenth century needed to keep readers coming back for the next weekly or monthly installment, they invented the device that eventually acquired the name *cliffhanger*. The English novelist Thomas Hardy was likely the first author to leave a character literally hanging from a cliff, at the end of Chapter XXI of *A Pair of Blue Eyes*, which ran in *Tinsley's Magazine* from September 1872 to July 1873. "Knight," Hardy concludes, describing his hero's relationship to the promontory, "was now literally suspended by his arms" — and there, suspended between life and death, Hardy leaves Knight, and us, hanging.

Hitchcock repeatedly exploited this visceral knowledge in his audience, even assigning a moral code to a character's relationship to heights: hero Cary Grant clinging to a cliff on Mount Rushmore in *North by Northwest* and managing to climb back up; villain Norman Lloyd hanging from the Statue of Liberty in *Saboteur*, then losing his grip and plummeting the three hundred feet to the harbor; morally compromised James Stewart at the end of *Vertigo*, standing on the ledge of a bell tower, staring down at the body of a victim of his obsessions, occupying a kind of midair purgatory: cured but coreless.

Denying gravity or defeating gravity is all very well and good if you're trying not to fall. If you're trying to rise, however, you need to do something at least equally heroic: You need to *defy* gravity.

When Laurel and Hardy try to push a piano crate up 131 steps, in their 1932 short *The Music Box*, only to have it roll back down, and then try to push it back up, only to have it roll back down, and so on, they're doing a riff on Sisyphus. The variations on the theme provide the specific laughs — the piano rolls over Ollie's back, Stan kicks a buggy-pushing nanny who mocks their misfortune,* a patrolman bops Stan in the noggin with a nightstick — but the theme itself is eternal.

What that theme is, is open to interpretation. Usually Sisyphus's task serves as a symbol of futility. Which it is — hence the adjective *Sisyphean.* But what else are we going to do, whether we're Sisyphus or Laurel and Hardy — *not* go up the hill? Consider the Sisyphean task within the context of defying gravity, and it becomes sort of noble — a symbol not of futility but of fortitude.

The narrative shorthand of falling and rising persists because whether we're watching a movie or walking on a sidewalk, the tension of being up in the air is real, and it can resolve itself only through a return to solid ground. Why is the riddle of the Sphinx — *What is the creature that walks on four legs in the morning, two legs at noon, and three in the evening?* — so enduring? Not just because the answer — *Man* — evokes the eternal truth of the aging process, but because of how it captures that truth. Its *Aha!* effect comes from our innate, unthinking understanding that life is one long rise and fall. But then, that's what stories are — one long rise and fall. We have embedded gravity in the narrative form itself: the rising and falling action of the dramatic arc.

And so we go, step after step, repeating a cycle of surrender and

---

* Nanny, to Patrolman: "And not only that, he kicked me!" Patrolman: "He kicked you?!?" Nanny: "Yes, Officer. Right in the middle of my daily duties." The prat also rises.

defiance. Of fear and fortitude. Of falling and not falling. Of maintaining our balance.

Just like the universe.

❖

I am the universe.

The theoretical physicist was standing at the front of a lecture hall at a meeting of the American Astronomical Society. He was perching on one leg, like a pelican, and holding his arms straight out to either side, like a human. He stayed perfectly still, or at least as still as his anatomy could manage. And as long as he stayed still, he could continue to claim that he was a likeness of the universe, within a reasonable margin of error.

But then he wasn't the universe anymore. He was just a guy with a proprioceptive system going into overdrive. His leg wobbled. His arms wavered. By shifting the muscles in his leg and the angles of his arms and the position of his hips, he was able to remain upright for a few more seconds. But then — no dice: He had to plant his other foot back on the floor. Either that or fall.

The point of his demonstration was to underscore how vanishingly improbable it is that the universe is in gravitational balance. Newton was right to ask why the universe hadn't collapsed, though his invocation of God as the answer was dubious. Einstein was right to ask, too, though his insertion of lambda to restore the balance of the universe was equally questionable. The eventual discovery that the universe is expanding, however, had removed the need for lambda, as well as the need to ask why the universe is in gravitational balance: It isn't! No worry!

In 1998, however, Einstein's lambda returned: Worry!

In the beginning has an obvious corollary: *In the end.* As the accumulating evidence from the cosmic microwave background began corroborating the Big Bang theory in the last quarter of the twentieth century, some cosmologists turned their attention to the other end of the spectrum of cosmic time. They assumed that an expanding universe full of matter that is interacting with all the other matter through the "force" of gravity is a universe that will not be expanding forever. It will either expand as far as it can, then collapse back on itself in a reverse Big Bang, or it will expand as far as it can, then coast for all eternity. In 1970 Hubble's protégé Allan Sandage argued in an era-defining paper that the future of cosmology rested on the discovery of "two key numbers." Just as you can predict when and where a moving car will stop if you know how fast it's going now and how much it's slowing down, so you can discover the fate of the universe if you know how fast it is currently expanding and how much that expansion is decelerating.

In 1998, however, two rival teams of observers reached the identical, counterintuitive conclusion: It's not decelerating. The expansion of the universe is not slowing down. It's speeding up.

Further observations showed that in the early universe the expansion had indeed been doing what observers would naïvely expect — slowing down. But around seven billion years ago the universe "turned over": The expansion began to accelerate. Einstein's lambda and Newton's *Pantokrator* were back.

In introducing lambda, Einstein had been trying to make his math *down here* match the motions *up there*, which he interpreted to be stable. Then the discovery of the expansion of the universe had seemed to eliminate the need for a stabilizing variable. The equation worked without it. The equation worked *with* it, too, if you set lambda to zero. The discovery of an expansion-*plus,* however, revived the need for a variable equal to non-zero in order to match the new vision of the universe.

Scientists named this mysterious anti-gravitational "force" *dark energy* to echo *dark matter*, another unnerving discovery of the late twentieth century. Throughout the 1970s astronomers had begun noticing that spiral galaxies were rotating in a way that didn't make sense. Our own spiral galaxy—the Milky Way—was rotating in a way that didn't make sense. Galaxies aren't solid objects; they're agglomerations of stars and gas and dust—loose fragments all whirlpooling as one. If a galaxy were a solid object, all the pieces would be turning together. But because it's made of loose matter, the material should be churning at rates that bear an inverse-square relationship to the center of the galaxy, just as the planets orbit the Sun at rates that bear an inverse-square relationship to the center of the solar system. Astronomers were observing, however, that the galaxies weren't following that law—and not just a galaxy here or there, but just about every spiral galaxy they examined. Either gravity as we know it varies in strength across the universe—a notion that would overturn the "universal" part of a universal law of gravitation—or some other kind of matter is gravitationally herding the gas and stars and dust together. Computer modeling showed that if you placed such a galaxy inside a cloud of "dark matter," as the mysterious substance was increasingly being called, it did in fact follow Newton's law.

Beginning around 1980, theorists began predicting what properties such matter should have — mass and so on — in order for it to behave the way it appeared to be behaving; experimentalists simply needed to go out and design the particle detectors to find that matter. Theorists and observers alike began forecasting that an answer to the question would be forthcoming within five years. Every five years, they made the same forecast. This pattern repeated itself until about 2010, by which point many physicists had learned their lesson and stopped predicting the impending identification of dark matter.

Nonetheless, the evidence was there, everywhere they looked — and not just in individual galaxies. As telescope technology improved — most notably, the many-order-of-magnitude increase of photon-collecting made possible by charge-coupled devices, or CCDs, in the 1980s — astrophysicists could peer deeper into the universe, which meant farther back in time, and they could see with far greater precision. What they found wasn't a random distribution of matter but a pattern: clusters of galaxies and, more important, superclusters of galaxies — garlands of galaxy clusters spanning anywhere from several hundred million to ten billion light-years. Separating these clusters of matter weren't non-clusters of free-floating matter but vast voids, as if something had swept the matter aside with a sidewalk broom. And in a way, something had: dark matter. Simulations showed that a stark disproportion between the accumulation of matter and the emptiness of space wouldn't be possible without expansive dark-matter gravitational exertions. Together the superclusters and empty space looked, on the largest scales, like a neural network or a cobweb or, for that matter, a skeleton.

The timing of these two discoveries — dark matter and dark energy — was not coincidental. One reason they happened at the same historical moment was that cosmology had become a science — a branch of physics, not metaphysics. Working on the scale of the universe, theorists could make predictions — for instance, that a cosmic microwave background, if it existed, would carry a certain frequency "visible" everywhere in the heavens — and observers could go out and validate them or not. The discoveries of an expansion, of dark matter that's dictating the evolution of the universe, of dark energy that's speeding up the expansion — of neutron stars and supernovae and black holes — had revealed the universe to be more violent and more volatile than we'd ever imagined, but also more varied. Sort of the opposite of the single-substance-rotating-in-circles perfection that the ancients imagined. But as had escaped the notice of no one working in the field, all these discoveries shared something other than timing: gravitation.

Gravity had always been an outlier in modern physics. Not just in the Newtonian sense that we don't understand the immediate cause for what we perceive as gravitational effects, but in the sense that as our understanding of the workings of the universe had grown, gravity continued to refuse to play well with others.

If gravity were merely the weakest force in the universe, we might shrug: Well, *one* of the four forces has to be weakest. (Of course, roping gravity into the "force" realm requires a bit of rhetorical legerdemain, but for now let's acknowledge that it is one of four processes that apparently hold the universe together.) But gravity is not only the weakest, it's the weakest *by far.*

The strong nuclear force, governing the stability of atomic nuclei, is around one hundred times stronger than the electromagnetic force, which in turn is up to ten thousand times stronger than the weak nuclear force that determines radioactive decay: three forces, all within a factor of one million of one another.

And then comes gravity. The weak nuclear force, the one that's a million times weaker than the strong nuclear force? It's about a million billion billion billion times stronger than gravitation.

Place a paper clip on a tabletop. There it remains, unmoving, anchored to its spot by the vast gravitational pull of the entire planet beneath it. Earth is an object with a mass of 6,583,003,100,000,000,000,000 tons. The paper clip is an object with a mass of 4/100 of an ounce. Now take a refrigerator magnet and wand it over the paper clip. Presto! You have counteracted the gravitational force of the entire Earth with a wave of your hand.

But not only is gravity bizarrely weaker than the other forces, it's the only one that doesn't have a quantum solution that incorporates Einstein's relativity. The strong nuclear has quantum chromodynamics, electromagnetism has quantum electrodynamics, the weak nuclear has quantum flavordynamics. Gravitation has quantum bupkisdynamics. The discovery of the graviton — a hypothetical particle that would mediate with nature on gravity's behalf in the same way that the gluon does for the strong nuclear, the photon does for electromagnetism, and the W and Z bosons do for the weak nuclear — would help. But if it exists, it has escaped detection via a cunning unparalleled in quantum experiments.

Aristotle, Newton, and Einstein had assumed the universe was stable because it looked stable on the scale of time familiar to each of them. In trying to understand the frightful balance of the twenty-first-century universe, cosmologists certainly were wary of making the same mistake as their predecessors. Besides, they actually knew the scale of cosmic time; the universe, as numerous observations (including cosmic microwave background explorers) had shown, is about 13.8 billion years old. So instead of trying to understand the gravitational balance in the universe by thinking about the scale of time, they turned to thinking about the scale of space. Instead of looking at the universe in the current moment and missing all the other moments, maybe we had been looking at the universe in its entirety and missing all the other universes.

That concept would explain a lot, not least because it would explain gravity. Not what gravity *is,* and not what *causes* gravity, but why gravity is the outlier it is.

In the beginning, the universe was nothing. In the quantum interpretation of nature, however, even nothing is something. It's potential — specifically, it is the potential to exist. Whether the nothing fulfills that potential is subject to probability. The odds are that nothing will come to nothing. But any one nothing might beat the odds and do something. If so, the something it does will be *to come into existence.*

This outcome is not a mathematical parlor trick — a theoretical possibility that's not a necessity. As Einstein — that longtime equivo-

cator on the existence of black holes — concluded in a lecture at Oxford in 1933, "Experience remains, of course, the sole criterion of the
physical unity of a mathematical construction." Not that you gotta
see it to believe it, but that if you see it, you gotta believe it.

And many physicists had seen it. Not directly, but in as convincing an indirect fashion as possible, which was as convincing as
much of modern physics got. In 1948, the Dutch physicist Hendrik
Casimir predicted that virtual particles would leave traces of energy.
Put two parallel plate conductors closer and closer together, and you
would be able to measure the increase in vacuum energy. Numerous experiments over the following decades validated the existence
of the "Casimir effect."

As odd as it is — "in tones of awe and reverence" was one mathematician's advice on how to approach the phenomenon — the Casimir effect holds a special meaning for gravity. According to the
general theory, energy interacts with gravity. And the Casimir effect
shows that virtual particles have energy. Therefore virtual particles
— bits of nothing that have managed to beat the odds and become
something — interact with gravity.

In the late 1970s and early 1980s, several theorists began exploring what this relationship might mean on a cosmological scale.
What they found in the math was that one trillionth of a trillionth of
a trillionth of a second after the universe came into existence, space
went through an "inflationary" stage that stretched its size a trillion-
fold. What they also found in the math is that if the universe did
arise out of a quantum pop — a nothing that became a something
— that pop almost necessarily would have created other pops. And
those other sudden somethings would, like the sudden something

that became our universe, become other universes. The most common mathematical interpretation placed the number of such universes, before the self-replicating mechanism shut off, at 10^{500} — a one followed by 500 zeros.*

If the inflationary scenario was a valid interpretation, then maybe the reason gravity is such an outlier that it may as well belong to a different universe is that it does. Theorists proposed, for instance, that gravity might be something that bleeds into our universe from an adjoining universe, or that it's an artifact from a colliding universe. Whatever the merits of individual theories about other universes, the idea of a multiverse as an explanation for gravitation's outlier status made the transition from uncommon to common unintelligibility within a single generation.

The overarching concept was called the anthropic principle. *Anthropic* means "relating to the existence of humans," and the principle, at least as it concerned cosmology, is that the reason we can examine the universe is that we live in a universe we can examine.

The logic was not as tautological as it might sound. In a multiverse, each universe would possess its own laws of physics. A universe in which, for instance, gravitation is not weaker than the strong nuclear force by a factor of about a million billion billion billion is a universe that would not be conducive to the existence of the human mind. Or to the existence of galaxies, for that matter.

---

* For comparison's sake: The universe — or at least *our* universe — contains 10^{80} atoms, which is a one followed by "only" 80 zeros; the photons that all the stars in the universe have emitted in the history of the universe number 4×10^{84}, or a four followed by 84 zeros.

"The history of astronomy," Edwin Hubble once wrote, "is a history of receding horizons." Aristotle's spheres defined our farthest view; then Galileo's congeries of stars within our galaxy; then Hubble's congeries of galaxies; then superclusters of galaxies, all hanging by a dark-matter thread; then a web of threads billowing from dark energy's breath. Why *not* other universes?

If you attended conferences and symposia relating to cosmology at the turn of the twenty-first century, you almost certainly saw a PowerPoint (or, more likely, an overhead projector) presentation on this topic. I once witnessed a theorist delivering a thunderously emphatic anti-anthropic lecture at a symposium at the Space Science Telescope Institute, on the Johns Hopkins University campus. Two or three years later I came across his byline above an essay in *Science* magazine in which he and a co-author argued that the anthropic principle was well worth exploring.

First the essay recounted the sort of anti-anthropic reasoning he'd presented at that lecture in Baltimore: "The potential existence of an ensemble of unobservable universes appears to be in conflict with the 'scientific method' (which requires theories to be falsifiable by observations or experiments) and therefore in the realm of metaphysics." Now, though, he wanted to make a distinction—"a 'fuzzy' boundary between what we define to be observable and what is not." We don't *see* dark matter or dark energy, but we know *something* is there. We don't *see* gravity, but we know something is there. You *don't* gotta see it to believe it, if what you can *see* leaves you with little or no alternative. In the case of anthropic reasoning, the deciding factor was gravity's strength, or lack thereof: "Were gravity not so weak, there would not be such a large difference between the atomic and the cosmic scales of mass, length, and time."

When I next saw him at a conference, I stopped him in a hallway, said I'd read his essay, and reminded him of the talk I'd attended. Then I said, "What happened?"

He shrugged. *Facts* was what happened. "The evidence," he said. Successors to the radio telescope that, in 1965, detected the cosmic microwave background had continued to refine our vision of the earliest universe: the Cosmic Background Explorer in the early 1990s, the Wilkinson Microwave Anisotropy Probe of the next decade, the Planck Observatory in the decade after that. By studying the quantum fluctuations in the primordial universe, physicists had quantified the distribution of matter and energy that was present then—and, the laws of conservation being what they are, that is present now. Those fluctuations revealed a universe that is 68.3 percent dark energy, 26.8 percent dark matter, and 4.9 percent ordinary matter (protons, neutrons, and so on—the stuff we'd always assumed was the universe in its entirety). For simplicity of presentation, those numbers had rounding errors, but if you dug into the quantum quantifications, the universe that emerged was in balance to an almost frightening degree. Dark energy, for instance, possessed the precise Planck density—a quantum measure that physicists use—of a decimal point followed by 122 zeros and then the number 136. If the value were even the least bit different, our universe wouldn't exist in any recognizable form. If, for instance, the value were a decimal point followed by 122 zeros and the number 137, galaxies wouldn't have formed. Yet form they did, as did we. What are the odds? One in 10^{500}?

This balance that is otherwise inexplicable, he said, left him with little choice: "It's very difficult *not* to have multiverses." We live in the universe in which gravity is extraordinarily weak because it's the only kind of universe out of which we could arise.

❖

My inbox was filling again. On February 11, 2016, at 10:31 a.m. ET, the email alert on my laptop pinged. It pinged again at 10:32, and then again at 10:45, and at 10:54, and at 10:56. By the end of the day I had received twenty-six press releases from various foundations, research centers, and universities announcing one of the most momentous discoveries in the history of physics: the detection of gravitational waves.

The news itself was not surprising. Online rumors had been rampant for days. The result — the actual detection of gravitational waves — wasn't surprising, either. If any experiment was going to detect gravitational waves, it was the Laser Interferometer Gravitational-Wave Observatory, or LIGO. Discovering gravitational waves was what its inventors had designed it to do; this latest iteration of LIGO's instruments would find them, if they were out there.

If.

Newton had assumed that gravitation, unlike light, acts instantaneously. Remove the Sun, he thought, and immediately Earth would adopt an inertial straight-ahead movement (plus a relatively minor centripetal interaction with the Moon and other planets). But the later discoveries of two unlikely coincidences challenged Newton's conclusion. One, the inverse-square law applies to both electricity and gravitation. Two, electricity propagates through waves. Therefore, perhaps gravitation propagates through waves as well — *ondes gravifiques*, as the French philosopher Henri Poincaré called them when he raised the possibility in 1905, while thinking about the kind of relativistic problems that Einstein was independently considering in his special theory of relativity. A decade later, Einstein applied that idea to his emerging general theory, but in the end he sort of concluded, as he wrote to Schwarzschild in February 1916, "there are no gravitational waves analogous to light waves."

Sort of because he later reconsidered without coming to a definitive conclusion, then reconsidered yet again without coming to a definitive conclusion, before coming to a definitive conclusion: His opinion didn't matter. Nobody was going to detect gravitational waves anyway. Such a detection would require a combination of observations at both extremes: the most powerful phenomena in the universe, aside from the origin event itself; the subtlest detection in history.

Advances in technology, though, have a way of easing the transition between uncommon and common unintelligibility. In the case of gravitational waves, that technology was the invention of the laser in 1960.

A laser differs from regular light in two important ways: Its focus is narrow (that is, it doesn't disperse in every direction, as "normal" light would), and its focus goes deep (that is, it stays narrow over long distances). As several researchers in the Soviet Union and the United States independently proposed, that combination — narrow, deep — might one day allow for measurements precise enough to detect gravitational waves, especially if the laser is beaming through an interferometer, a device that simultaneously sends two light signals down different paths of the same length and measures their arrival times at a common destination. The arrival times will be simultaneous unless something has interfered with the two light waves during their separate journeys. At least in principle, the researchers argued, gravitational waves could be that something. These ripples in space-time would affect the atoms in the arms of the interferometer, changing the length of the arms and therefore the travel times for the light.

Among the US researchers advocating that approach was Rainer Weiss, a physicist at the Massachusetts Institute of Technology. By 1972 he had completed the first gravitational-wave interferometer that could account for the "noise" in the signal-to-noise ratio — a crucial consideration when measuring the effects of a cause as pervasive yet subtle as gravitation.

During the same period, Kip Thorne founded a research group at the California Institute of Technology to study the theoretical side of gravitational waves. When Thorne asked Weiss what kind of technology would complement the theoretical work, Weiss convinced him that interferometry at a large enough scale actually had a realistic chance of detecting gravitational waves.

LIGO would consist of two interferometers, in part to guard against a single false-positive detection; one instrument would be in Louisiana, the other in Washington State. At each location, lasers would travel back and forth along two corridors, each corridor two and a half miles long — long enough that the designers would have to compensate for the curvature of the Earth. Even that distance, however, wouldn't be enough to detect something as sensitive as a gravitational wave. Instead, mirrors at each end of each corridor would bounce the laser beam back and forth, back and forth, hundreds of times, until the effective distance that each beam covered would be just shy of seven hundred miles.

At that point, LIGO would be able to reach its goal of detecting a difference between the two arms of the interferometer made by a passing gravitational wave: one ten-thousandth the width of a proton.

Scientists often rely on mind-bending metaphors to impress upon the public the scale of their experiments — for instance, a telescope so sensitive it could see a golf ball on the Moon. To find the proper metaphorical scale for LIGO, however, the scientists had to leave the solar system. One ten-thousandth the width of a proton is to the two and a half miles of an interferometer corridor as a human hair is to the distance between the Sun and the next-nearest star, Proxima Centauri, 4.243 light-years, or 25 trillion miles, away. (And to try to understand *that* scale, consider this: To count to 25 trillion at the rate of one number per second would take 791,000 years.)

The LIGO team knew the first iteration of the instrument wasn't going to make that detection; it would provide a dry run of sorts to test the fundamentals of the technology. The second round of observations would follow a significant technological upgrade. "You get the apparatus to a certain point," Weiss once told me, "and then you just put your hands in your pockets and wait."

Or not wait. The detection itself might have been unsurprising, but not the rapidity with which the new instrument made it: The updated LIGO wasn't even on line yet, at least officially. It was still in the testing stage. That the detection happened so quickly raised the question: Was the LIGO team some kind of crazy lucky, or was the universe spewing out gravitational waves of a certain magnitude —large enough to be "visible" to a detector on Earth—at such a rate that a nearly immediate detection was almost inevitable?

The detection came with a second surprise: the nature of the source. Of the various gravitational-wave producers that LIGO might observe—the kinds that disturb space-time to such an extent that LIGO could register the aftershock—the collision of two black holes orbiting each other was perhaps the least likely. Exploding stars, neutron stars, colliding neutron stars: These were what the LIGO collaboration foresaw as far more common candidates. The fact that the first detection was of a phenomenon that the scientists had expected to be the rarest raised another question: Are collisions between orbiting black holes more frequent in the universe than we assumed?

Seems so. Only three months later, on December 26, 2015, LIGO made a second detection — another pair of colliding black holes. A year later, another pair of black holes. Six months later, yet another pair of black holes. A month later, still another. Then, three days later, on August 17, 2017 — *at last:* a collision between neutron stars, the phenomenon that the LIGO collaboration had thought would be among the most frequent.

Scientists love surprises. They want answers to their questions, and they design experiments in order to answer their questions, but they want new questions, too. In this case, the unanticipated question was, Just how frequent are black-hole collisions in the universe? Extrapolating from the existing data, theorists concluded that binary black holes collide somewhere in the universe every five minutes. Common unintelligibility, indeed.

And here's what's common: one black hole, ten million times the mass of the Earth, orbiting another black hole, also ten million times the mass of the Earth, at a distance of 125 miles, at the rate of 300 times per second.

This information — the frequency of the collisions and what (we're pretty sure) they actually entail — might have seemed to add another barrier between us and the universe. Probably, it did. But you know the saying: Two steps forward, one step back. Even one step forward, one step back would be impressive, given how formidable each step is, whether our own or the universe's.

For astronomers and physicists, the detection of gravitational waves is one step forward. But just as significant as information about individual phenomena is the next step: what gravitational waves portend for astronomy and physics themselves. For thousands of years we studied the sky with our eyes alone. Beginning with Galileo we studied the sky with a technology that vastly improved our vision. In the mid-twentieth century we began studying the sky in areas of the electromagnetic spectrum that our eyes couldn't see, with or without the aid of an optical telescope. Now, we could begin to study the sky without light of any kind at all. Instead, we could use distortions in space-time — gravitational waves emanating not just from black holes or neutron stars but from the Big Bang. That kind of breakthrough elicits from scientists their favorite question of all: What's next?

❖

And so the ancient conversation continues.

The current scientific understanding of the universe is not a creation myth, a collection of fantasies to help explain what this place is, and what our place is within it. Instead, it's an origin story — a collection of facts that our minds have shaped into a narrative. It's also incomplete, an equation with an x we will continue to try to solve, or at least try to solve as much as we can, generation by generation, if only because what choice do we have — *not* to push the piano up the hill? Not to pile mountain upon mountain or to fashion wings from feathers — or perhaps even to inscribe boulders that lie off the beaten path? What kind of species would we be if we curtailed our ambitions simply because *Oh, the climb is hard* or *Oh, the fall is far*?

That's all we're doing, is rising and falling. We wake up, we face the day, we retire. We crawl, we walk, we lean on a cane. We sit, we stand, we fall — though not usually. Usually gravity simply wears down our resolve. And then, again, we bring our body as close as possible to the earth. We sit, or lie, or kneel.

But sometimes we remember to lift our eyes long enough to see what temple we have entered. Only then, and only if we're fortunate, might we receive the reward of a glimpse of a unity that would otherwise elude us — an *equivalence,* you might call it, if you're of an Einsteinian frame of mind: between inertial motion and centripetal motion, between space and time, between *up there* and *down here,* between *In the beginning* and *The end.*

GRAVITY:
IN CONCLUSION

Pompeii. The present.

Our modest tour group—my wife and I, along with another middle-aged American couple we met this week at our hotel—is standing at the intersection of Via di Mercurio and Via della Fortuna. The guide we've hired is trying to impress upon us the similarities between ancient civilization and our own.

"Look!" he says, and he points to a frieze at the top of a nearby building. We look. The figures are inscrutable at first, but then the guide explains: The building had been a shop belonging to a wine merchant. *Ahhh,* we go, not at the shop's having belonged to a wine merchant but at the utility of the sign. To anyone who was familiar with the sight of two men hoisting a wine bladder—as visitors from throughout the Mediterranean were, back when this establishment was a going concern — the identity of the shop would have been at-a-glance unmistakable.

"Look!": pointing now to a metal pipe sticking out of the soil abutting the base of a house. We look. Our guide explains: "Indoor plumbing. Two thousand years ago!"

"Look!": pointing to a vestibule floor where a mosaic reads CAVE CANEM. We look. "'Beware of dog.' Two thousand years ago!"

We hadn't known what to expect from Pompeii; none of us in our group had known, we would learn later, over dinner at our hotel in Positano, an hour from here by car. But what we are finding is that our guide is correct: Pompeii now is, essentially, what Pompeii was then, and what Pompeii was then was, more or less, what civilization is now and has always been: corner shops; homes with frescoes and indoor plumbing and welcome-mat warnings; roads with chariot ruts — and all of this familiar life stretching, block after block after intact block, to the horizon.

The horizon here, at this intersection just off the main square in the heart of the city, is Mount Vesuvius. The city planners presumably made sure the main intersections afforded this view. It's impressive now, this view, but it must have been far more impressive then, before the volcano lost its upper two-thirds.

I have made the day trip to Pompeii specifically to stand in a spot like this and to consider a view like that and to try to lift my eyes long enough to see what temple I have entered. I'm hoping for a glimpse of a unity, an equivalence, not just between the *then and there* and the *here and now* — the ancient world and ours — but between the *up there* and the *down here,* the universal and the individual. When nobody is looking, I pull my smartphone out of my pocket and tap a GPS app. GPS technology wouldn't work if its designers didn't take into account the warping of the universe predicted by the general theory of relativity. I don't care what the co-ordinates actually are; I just want to establish a connection between the Einsteinian space-time continuum and the Aristotelian spheres. That I'm doing so at a site where a philosophically minded soul — two thousand years ago! — might have thought that the Bastarnae had been right and that the sky does fall, should be a bonus. But I'm only human: I can't get there from here. I observe the coordinates on the app without absorbing them, pocket my smartphone, and catch up with the tour.

The four of us are among the guests at a writing conference, and that evening we join the rest of the attendees at the hotel res-taurant. The hotel is on a hillside, as is everything in Positano — as is pretty much everything along the Amalfi Coast that isn't actually on a beach. To my right I can see the lights of the houses at the top of the hill; the wealthy in this region have always lived there, in part because, in an earlier age, gravity would have exhausted any seafar-ing conquerors before they reached the summit, and in part because it's the real estate with the most glorious view. To my left, at the base of the hill, is the view: the Tyrrhenian Sea, though at the moment it's mostly invisible. All I can see of it is the lip of it breaching a concrete pier lit by a blue light.

The sea is wild tonight. The surf easily overruns the beach, rearing up like an asp to strike second-story windows that are usu-ally out of reach. To one side of the dining room the wait staff, locals who have spent a lifetime here, are whispering among them-selves: They've never seen anything like it. From this God's-eye perspective, one could easily imagine the world not as a stationary globe, and not even as a spinning globe, but as a spinning globule: a giant drop of water, sloshing. I do, anyway. I imagine that the water "wants" to go straight, and that it "thinks" it is going straight, and that it's "correct" in thinking that it's going straight, at least from its own perspective, but the waves also seem to be straining toward the Moon and angling toward the Earth, which they are, from a Newtonian perspective, whereas from an Einsteinian perspective those two Newtonian motions are achieving a synthesis, an *equivalence* —

"You're fading," the woman to my right says. She has asked me something, I realize, and I, looking past her to the sea, haven't noticed.

"No, I'm here," I say. "I was just thinking."

Earlier I had explained to her the idea behind the book I'm writing: that nobody knows what gravity is, and almost nobody knows that nobody knows what gravity is except for scientists, and they know that nobody knows what gravity is because they know that *they* don't know what gravity is.

Now she follows my gaze, over her shoulder and toward the sea, and then she turns back to me. She's smiling as if we share a secret.

"You must see gravity everywhere," she says. Then she tells me a story about her son when he was young. He was only two or three years old, she says. He was sitting in his high chair, and he bumped his sippy cup overboard. On purpose, he bumped it. He knew what would happen, and he wanted it to happen, but he wanted it to happen so he could watch it happen. "And I said to him, 'What made it fall?' And he said, 'Gravity.'"

I laugh. And then, because I can't help myself: "Did you ask him if he knows what gravity is?"

❖

"What is gravity?"

This time I'm asking the question of five physicists at once, as the moderator of a panel at the World Science Festival in New York in June 2016. The announcement of the discovery of gravitational waves isn't even four months old, yet the schedule at the festival is packed with events about gravitation. The festival has given my panel the punning title "Getting a Grip on Gravity," positioning it as an accessible alternative to the more science-centric presentations. I open the discussion by telling the panelists that I once asked Kip Thorne the same question, and that I will tell them Kip Thorne's answer after they've given me theirs.

Ten seconds of silence might not seem like a long time, but on-stage it's an eternity. Fine by me. I want the audience to think about the question like they've never thought about the question before — to realize, perhaps, that they've never thought about the question before.

Finally, David Gross volunteers. As the Nobel laureate in the group (Physics, 2004), he probably feels a responsibility to help the discussion along. He explains to the audience the two great theories in physics and why they are so far irreconcilable. Despite this so-far-irreconcilability, he says, if we want to send a rocket to the Moon or calculate how much a gravity wave might shrink or stretch the arms of a LIGO outpost, we can do so. "But *What is?* for somebody like me," he says, "raises question marks and not the answers that we already know."

"Right," I say. Then I reveal to the panel Kip Thorne's answer — that *What is gravity?* is a meaningless question.

Gross laughs. Then another panelist, a Caltech particle physicist, jumps in. She says she agrees with Gross that the question raises only further questions. And the lack of any answers, she says, is precisely what makes the discussion valuable. "So," she says, with emphasis, "I don't think it's a meaningless question."

Good enough. You can learn a lot from not-meaningless questions. I could, anyway.

Late in the evening of that dinner in Positano, back in our room, I stepped out to the veranda. The sea was, if anything, wilder now. I watched as the Earth seemed to shudder in its labors and the waves moved in both straight lines and downward curves of increasing disparity (and, when those actions met the equal and opposite reaction of immovable buildings, upward surges), and then I saw it. I glimpsed the equivalence that had twice eluded me that day.

Just as Galileo realized that if you drop a rock from the mast of a ship traveling on a river, its trajectory will appear to be an angle to an observer on the shore, so Einstein realized that the same principle applies to Newton's tides. The straight lines and downward curves (and upward surges) will not appear straight or downward (or upward) to an observer on a "shore" outside the Earth because from that perspective the actions of the ocean are part of a planet that rotates about its axis and moves around a star and travels together with the star toward the constellation of Hercules. But that much I already knew.

I had fallen again, a week earlier, at the beginning of our stay, this time while stepping out of the shower. The floor was marble; I hadn't put down the mat. My fall was over so fast I could only lie on the tiles, blinking and bleeding. The sense memory has accompanied me all week, like the swelling of my elbow, and now it has followed me back from Pompeii: One moment you're strolling into the local wine shop or stepping out of a hotel shower, and the next, you're not. But that much I already knew, too.

What I hadn't fully appreciated in my years-long odyssey from bookstore chair to bathroom floor was that the same Einsteinian principle — the erasure of the distinction between *down here* and *up there* — applies to me. And then some: unlike Einstein, I knew that my rotating, orbiting, Hercules-heading planet is also revolving around the center of a galaxy that's moving toward other galaxies that belong to a universe that's undergoing an expansion that's accelerating. To me? I fell down. To you, on a distant shore? My position changed in relation to the Earth's, and the Earth's position changed in relation to mine, and you and I will say these motions are the effect of something we can agree to call gravity.

Small comfort for me and my swollen elbow down here, but in this universe, you take what you can get.

I still didn't know what gravity is, but when I closed the veranda shutters and turned in the darkness toward my bed, I no longer minded. At least I was beginning to suspect why anyone would ask the question. It would be for the same reason that we chart the planets and reach for the stars, that we study birds and follow the Sun, that we eye the horizon: because each of us sleeps in the shadow of our own Vesuvius, and there we dream we can fly.

ACKNOWLEDGMENTS

Deepest appreciation to editor Alexander Littlefield for his immediate enthusiasm and his ongoing guidance, patience, and advice. Special thanks to Olivia Bartz for amplifying Alex's insights and adding her own valuable contributions. As always, gratitude to Henry Dunow, agent for lo these many years. Awe, again, at Katya Rice's embodiment of copyeditor as collaborator. Thanks to Courtney Young for shepherding and shaping the idea for this book, bouquets to Bob Morris for the title, bows to Ron Cowen, Laura M. Mac Donald, Greg Hollingshead, and Nicholas Suntzeff for essential advice. Gratitude to the Faculty Development Fund at Goddard College. And love to Gabriel and Charlie, who are now out of the house and therefore not in need of apologies for what would have been, yet again, their father's neglect as he finishes a book. (Those apologies go instead to the author's wife, who is also this book's dedicatee.)

NOTES

CHAPTER 1

PAGE

9 *Just ask the Celts:* Quotations from creation myths come from Sproul.

11 *"the primeval pair":* Rose, p. 20.

16 *Palaephatus:* Palaephatus.

17 *Er explains:* Bloom and Stewart are the primary sources for the myth of Er.

21 *the highest point:* Aslan, p. 6.

22 *Herakles was burning:* Ovid.
    conversation with Diotima: Plato, *Symposium.*

23 *Plato often advocated:* Crowe, p. 23.
    Eudoxus of Cnidus: Ibid., p. 24.

25 *the torments of the flesh:* Gardiner is the primary source for these Christian visions.

27 *Vibia Perpetua:* Ibid., p. 133.
    The future saint Balthild: Ibid., p. 135.
    The future saints: Ibid.

31 *Commedia:* Dante 1939.

32 *Convivio:* Dante 1990.
33 *"from tuft to tuft":* Dante 1939, pp. 425–427.

CHAPTER 2

35 *"The ancients":* Aristotle.
36 *"The inquiry into nature":* Ibid.
The Greek poet: Cook, p. 55.
When the Roman playwright: Ibid.
When the Celts: Ibid.
37 *Livy recounts:* Livy.
One Egyptian myth: Cook, p. 126.
The Cherokee proposed: Sproul, p. 254.
The Greeks split: Keightley, p. 29.
according to Hesiod's Theogony: Hesiod, p. 131.
38 *every "old tale":* Aristotle.
44 *He and his fellow students:* Sorabji 2012, p. 1.
Philoponus being a nickname: Sorabji 1987c, p. 5.
46 *To the traditional Greek mind:* Ibid., p. 6.
one of the few: Sorabji 2012, p. 8.
47 *one of them composed:* Sorabji 1987c, p. 2.
47 *Philoponus provided it:* Sorabji 1987b, p. 168.
Philoponus, however, realized: Ibid., p. 171.
48 *The number of times:* Sorabji 1987a, p. 20.
49 *generation or destruction:* Sorabji 1987c, p. 24.
though they could continue: Sorabji 1987b, p. 171.
"The Grammarian": Chase, p. 3.
Simplicius considered: Hoffmann, pp. 64–65.
50 *"I have tumbled":* Ibid., p. 69.
Simplicius was clinging: Ibid., pp. 66–67.
51 *the relics of martyrs:* Chase, pp. 3–4.
Thanks in part: Hoffmann, p. 63.
58 *"How," he wondered:* Rosen, p. xvii.

CHAPTER 3

66 *The latter infractions:* Westfall, p. 77.
the other notebook: Ibid., p. 89.
67 *"Of Gravity & Levity":* Newton 2003.
68 *"Gummy," he guessed:* Dobbs, p. 102.
69 *a fifty-two-page pamphlet:* Galileo 1989.

72 *Newton agreed:* For the trajectory of Newton's thoughts, see especially Cohen, Gleick, Westfall.
    "Kepler," Newton wrote: Smith.
    "the language of mathematics": Galileo 1957, pp. 237–238.
74 *In the summer of 1684:* Newton 1978, p. 237.
76 *No and no:* For the trajectory of Galileo's thoughts, see, e.g., Westfall, pp. 6–10.
79 *"These things":* Philoponus 2012, p. 41.
80 *Imagine you're on a ship:* Galileo 1967, pp. 139–145, 186–187.
82 *"Since the Bible":* T. Kuhn, p. 121.
84 *"Therefore the major planets":* Newton 1978, p. 277.
85 *They aren't points:* Cohen, p. 18.
    "The planets neither": Newton 1978, p. 281.

CHAPTER 4

90 *"I am not especially":* Smith.
91 *"It is true":* Leibniz.
    "That Gravity": Newton 1756, pp. 25–26.
92 *Toward the end:* Newton 1999, p. 399.
96 *"Now something":* Ibid., p. 385.
    "In the discovery": Gribbin, p. 71.
97 *"There remains":* Bacon 2001.
100 *"whose natural philosophy":* Newton 1999, p. 386.
    Newton introduced: Ibid., p. 794.
102 *"I have not as yet":* Ibid., p. 943.
104 *His book opened:* Gordon, p. 1.
105 *In the introduction:* Pirrie, p. ix.
    One such test: The primary source for the Cassini expeditions is Ferreiro.
107 *Newton used Richer's:* Newton 1999, p. 829.
109 *In 1701, a team:* Ferreiro.
110 *"England may justly":* Sprat, pp. 113–115.
111 *as Newton predicted:* Ferreiro.
115 *"A Frenchman":* Voltaire.
    Voltaire was going to add: Arianrhod, chapter 4.
    The enterprise: Ibid., chapter 5.
116 *Not so the data:* Ibid.
118 *"I am more":* Broughton, p. 124.
119 *"Comets hav[e] of late":* Waff, p. 9.
    Alexis-Claude Clairaut: Broughton, pp. 126–127.
120 *As Halley had written:* Wallis, p. 281.

121 *"I cannot but"*: Ibid., p. 283.
"It cannot but": Ibid., p. 284.

CHAPTER 5

124 *On one side:* Laplace 1902.
"Let us recall": Ibid.
125 *Three decades later:* Waff, p. 7.
"would undoubtedly": Ibid., p. 4.
126 *Wesley promised:* Ibid., p. 5.
"As the Almighty": Ibid., p. 12.
127 *"a further knowledge"*: Bacon 1901, p. 42.
130 *A monotype:* Dobbs, p. 242.
"The souls of 500": These several examples come from Dawkins, pp. 38–40.
131 *"When I wrote"*: Newton 1756, p. 1.
132 *"They will indeed"*: Newton 1999, p. 940.
134 *"arrive without"*: Laplace 1902, p. 1.
135 *"a large Field"*: Halley, p. 22.
The following year: Broughton, p. 130.
"So far from": Laplace 1809, p. 3.
136 *"There still remains"*: Ibid., p. 372.
140 The Age of Reason: Paine.
"bring the world": Ibid., p. 1.
"We ought to regard": Laplace 1902.
143 *an 1872 book:* Mach 1898a, p. 57.

CHAPTER 6

145 *one-woman vigil:* West.
146 *"In glancing over"*: Mach 1898a, pp. 227–228.
147 *Einstein did experience:* Einstein 2002b, p. 136.
148 *"On the Electrodynamics"*: Einstein 1989, p. 140.
"possess no properties": Ibid.
153 *"The happiest thought"*: Einstein 2002b, p. 136.
154 *"Two eruptions"*: Einstein 2002a, p. 1.
"a giant vessel": Ibid.
156 *Einstein recruited:* Fölsing, p. 314.
"child's play": Ibid., p. 315.
165 *Georges Lemaître:* Block.
the same conclusion: Perlmutter et al.; Riess et al. 1998.
166 *One carried:* Dicke et al.

In the other paper: Penzias and Wilson.

173 *"It came in 1979":* Walsh et al.

theorist of his day: See, e.g., Thorne.

CHAPTER 7

180 *newly knighted PhD:* private video by Sabancilar Eray, courtesy of Alexander Vilenkin.

181 *"Gravity — Our Enemy":* Babson, p. 828.

182 *The review of:* Rose C. Feld, "Roger Babson's Life," *New York Times*, March 1, 1936.

"coming about": Gardner, p. 92.

Martin Gardner: Ibid., pp. 92–100.

186 *As a species:* Haldane, pp. 18–26.

The incremental anatomical: Hutson and Ward, pp. 163–170.

189 *Hitchcock delineated:* https://www.youtube.com/watch?v=OvNla9-u6xM.

191 The Heckling Hare: https://www.dailymotion.com/video/x2mdlfj.

195 *era-defining paper:* Sandage.

Further observations: Riess et al. 2001.

196 *Computer modeling:* See, e.g., Ostriker and Peebles.

201 *"Experience remains":* Einstein 1954, p. 274.

"in tones of": Fulling, p. 114.

several theorists began: See, e.g., Guth.

203 *above an essay:* Livio and Rees, p. 1022.

204 *the number 136:* Harman, p. 31.

206 *wrote to Schwarzschild:* Einstein 1998, p. 196.

210 *every five minutes:* Impey, p. xix.

what's common: Ibid., p. 195.

BIBLIOGRAPHY

Arianrhod, Robyn. *Seduced by Logic: Émilie du Châtelet, Mary Somerville, and the Newtonian Revolution.* Oxford: Oxford University Press, 2012.

Aristotle. *On the Heavens.* Translated by A. L. Stocks. http://classics.mit.edu/Aristotle/heavens.html.

Aslan, Reza. *Zealot: The Life and Times of Jesus of Nazareth.* New York: Random House, 2013.

Babson, Roger. "Gravity—Our Enemy Number One." In *Gravity's Shadow: The Search for Gravitational Waves*, by Harry Collins. Chicago: University of Chicago Press, 2004.

Bacon, Francis. *The Advancement of Learning*, edited by Joseph Devey. New York: P. F. Collier & Son, 1901. http://oll.libertyfund.org/titles/1433.

———. *Prefaces and Prologues*, vol. 39. New York: P. F. Collier & Son, 1909–14; Bartleby.com, 2001. www.bartleby.com/39/.

Berardelli, Phil. "Time Before Time." *Science Now,* July 5, 2007, p. 4.

Block, David L. "Georges Lemaître and Stigler's Law of Eponymy." In *Georges Lemaître: Life, Science and Legacy,* edited by Rodney D. Holder and Simon Mitton. New York: Springer Heidelberg, 2012.

Bloom, Allan. *The Republic of Plato.* Translated with notes and an interpretative essay by Allan Bloom. New York: Basic Books, 1991.

Broughton, Peter. "The First Predicted Return of Comet Halley." *Journal for the History of Astronomy* 16 (1985): 124–133.

Chase, Michael. "Simplicius' Response to Philoponus' Attacks on Aristotle's *Physics* 8.1." In *Simplicius: On Aristotle Physics 8.1–5,* edited by Richard Sorabji. London: Bloomsbury Academic, 2012.

Cohen, I. Bernard. "A Guide to Newton's *Principia.*" In *The Principia: Mathematical Principles of Natural Philosophy.* Translated by I. Bernard Cohen and Anne Whitman. Berkeley: University of California Press, 1999.

Collins, Harry. *Gravity's Shadow: The Search for Gravitational Waves.* Chicago: University of Chicago Press, 2004.

Cook, Arthur Bernard. *Zeus: A Study in Ancient Religion,* vol. 2, part 1. Cambridge: Cambridge University Press, 1925.

Crowe, Michael J. *Theories of the World from Antiquity to the Copernican Revolution.* New York: Dover Publications, 1990.

Dante Alighieri. *Dante's Inferno.* Translated by John D. Sinclair. New York: Oxford University Press, 1939.

———. *Il Convivio (The Banquet).* Translated by Richard H. Lansing. Abingdon, UK: Routledge, 1990. https://digitaldante.columbia.edu/text/library/the-convivio/.

Dawkins, Richard. *Unweaving the Rainbow: Science, Delusion and the Appetite for Wonder.* Boston: Mariner Books, 2000.

Dicke, R. H., P. J. E. Peebles, P. G. Roll, and D. T. Wilkinson. "Cosmic Black-Body Radiation." *Astrophysical Journal* 142 (1965): 414–419.

Dobbs, Betty Jo Teeter. *The Janus Faces of Genius: The Role of Alchemy in Newton's Thought.* Cambridge: Cambridge University Press, 2002.

Dorman, Thomas. "The Fascioligamentous Organ." In *Oxford Textbook of Musculoskeletal Medicine,* edited by Michael Hutson and Adam Ward. Oxford: Oxford University Press, 2016.

Einstein, Albert. "On the Method of Theoretical Physics." In *Ideas and Opinions.* New York: Crown Publishers, 1954.

———. "On the Electrodynamics of Moving Bodies." In *The Collected Papers of Albert Einstein,* vol. 2, *The Swiss Years: Writings, 1900–1909.* Translated by Anna Beck. Princeton, NJ: Princeton University Press, 1989.

———. "To Karl Schwarzschild." In *The Collected Papers of Albert Einstein,* vol. 8, *The Berlin Years: Correspondence, 1914–1918.* Translated by Ann M. Hentschel. Princeton, NJ: Princeton University Press, 1998.

———. "The Principal Ideas of the Theory of Relativity." In *The Collected Papers of Albert Einstein,* vol. 7, *The Berlin Years: Writings, 1918–1921.* Translated by Alfred Engel. Princeton, NJ: Princeton University Press, 2002a.

————. "Fundamental Ideas and Methods of the Theory of Relativity, Presented in Their Development." In *The Collected Papers of Albert Einstein,* vol. 7, *The Berlin Years: Writings, 1918–1921.* Translated by Alfred Engel. Princeton, NJ: Princeton University Press, 2002b.

Ferreiro, Larrie D. *Measure of the Earth: The Enlightenment Expedition That Reshaped Our World.* New York: Basic Books, 2011.

Fölsing, Albrecht. *Albert Einstein.* New York: Penguin Books, 1998.

Friesen, John. "Hutchinsonianism and the Newtonian Enlightenment." *Centaurus* 48 (2006): 40–49.

Fulling, Stephen A. *Aspects of Quantum Field Theory in Curved Space-Time.* Cambridge: Cambridge University Press, 1996.

Galileo Galilei. *Discoveries and Opinions of Galileo.* Translated and with an introduction and notes by Stillman Drake. New York: Anchor Books, 1957.

————. *Dialogue Concerning the Two Chief World Systems — Ptolemaic and Copernican.* Translated by Stillman Drake. Berkeley: University of California Press, 1967.

————. *Sidereus Nuncius.* Translated by Albert Van Helden. Chicago: University of Chicago Press, 1989.

Gardiner, Eileen, ed. *Visions of Heaven and Hell Before Dante.* New York: Italica Press, 1989.

Gardner, Martin. *Fads and Fallacies in the Name of Science.* New York: Dover Publications, 1957.

Gleick, James. *Isaac Newton.* New York: Vintage Books, 2004.

Gordon, George. *Remarks upon the Newtonian Philosophy.* London: W.W., 1719.

Gribbin, John. *The Scientists: A History of Science Told Through the Lives of Its Greatest Inventors.* New York: Random House, 2002.

Guth, Alan. *The Inflationary Universe: The Quest for a New Theory of Cosmic Origins.* New York: Basic Books, 1998.

Haldane, J. B. S. *Possible Worlds and Other Essays.* London: Chatto and Windus, 1927.

Halley, Edmond. *Synopsis of the Astronomy of Comets.* London: John Senex, 1705.

Harman, Oren. *Evolutions: Fifteen Myths That Explain Our World.* New York: Farrar, Straus & Giroux, 2018.

Hesiod. *The Homeric Hymns and Homerica.* Translated by Hugh G. Evelyn-White. New York: G. P. Putnam's Sons, 1920. http://www.sacred-texts.com/cla/hesiod/theogony.htm.

Hoffmann, Philippe. "Simplicius' Polemics." In *Philoponus and the Rejection of Aristotelian Science,* edited by Richard Sorabji. Ithaca, NY: Cornell University Press, 1987.

Holder, Rodney D., and Simon Mitton, eds. *Georges Lemaître: Life, Science and Legacy.* New York: Springer Heidelberg, 2012.

Hutson, Michael, and Adam Ward, eds. *Oxford Textbook of Musculoskeletal Medicine.* Oxford: Oxford University Press, 2016.

Impey, Chris. *Einstein's Monsters: The Life and Times of Black Holes.* New York: W. W. Norton, 2019.

Keightley, Thomas. *Classical Mythology: The Myths of Ancient Greece and Ancient Italy.* Revised and edited by L. Schmitz. Chicago: Ares Publishers, 1976.

Kepler, Johannes. *Kepler's Somnium: The Dream, or Posthumous Work on Lunar Astronomy.* Translated, with a commentary, by Edward Rosen. Mineola, NY: Dover Publications, 2003.

Kuhn, Albert J. "Glory or Gravity: Hutchinson vs. Newton." *Journal of the History of Ideas* 22 (1961): 303–322.

Kuhn, Thomas S. *The Copernican Revolution: Planetary Astronomy in the Development of Western Thought.* Cambridge, MA: Harvard University Press, 1957.

Laplace, Pierre-Simon. *The System of the World.* Translated by J. Pond. London: Richard Phillips, 1809.

———. *A Philosophical Essay on Probabilities.* Translated by Frederick Wilson Truscott and Frederick Lincoln Emory. London: John Wiley & Sons, 1902.

Leibniz, G. W. *Theodicy: Essays on the Goodness of God, the Freedom of Man and the Origin of Evil.* Translated by E. M. Huggard. Peru, IL: Open Court Publishing Co., 1985. http://www.gutenberg.org/files/17147/17147-h/17147-h.htm.

Livio, Mario, and Martin J. Rees. "Anthropic Reasoning." *Science* 309 (2005): 1022–1023.

Livy. *History of Rome: Books 40 to 42.* Translated by Evan T. Sage and Alfred C. Schlesinger. Cambridge, MA: Harvard University Press, 1938. http://www.hup.harvard.edu/catalog.php?isbn=9780674993662.

Luminet, J.-P. "Image of a Spherical Black Hole with Thin Accretion Disk." *Astronomy and Astrophysics* 75 (1979): 228–235.

Mach, Ernst. "On Transformation and Adaptation in Scientific Thought." In *Popular Scientific Lectures.* Translated by T. J. McCormack. Chicago: Open Court Publishing Co., 1898a.

———. *Popular Scientific Lectures.* Translated by T. J. McCormack. Chicago: Open Court Publishing Co., 1898b.

———. *History and Root of the Principle of the Conservation of Energy.* Translated by Philip E. B. Jourdain. Chicago: Open Court Publishing Co., 1911.

Markos, Louis. *Heaven and Hell: Visions of the Afterlife in the Western Poetic Tradition.* Eugene, OR: Cascade Books, 2013.

Newton, Isaac. *Four Letters from Sir Isaac Newton to Doctor Bentley, Containing Some Arguments in Proof of a Deity.* London: R. and J. Dodsley, 1756.

———. *Unpublished Scientific Papers of Isaac Newton.* Chosen, edited, and translated by A. Rupert Hall and Marie Boas Hall. Cambridge: Cambridge University Press, 1978.

———. *The Principia: Mathematical Principles of Natural Philosophy.* Translated by I. Bernard Cohen and Anne Whitman. Berkeley: University of California Press, 1999.

———. "Quæstiones quædam Philosophiæ." Cambridge: Cambridge University Library, 2003. http://www.newtonproject.ox.ac.uk/view/texts/normalized/THEM00092.

Ostriker, J. P., and P. J. E. Peebles. "A Numerical Study of the Stability of Flattened Galaxies: or, Can Cold Galaxies Survive?" *Astrophysical Journal* 186 (1973): 467–480.

Ovid. *Metamorphoses.* Translated by Anthony S. Kline. http://ovid.lib.virginia.edu/trans/Metamorph9.htm.

Paine, Thomas. *The Age of Reason.* London: Freethought Publishing Co., 1880. https://openlibrary.org/works/OL60357W/The_Age_of_Reason.

Palaephatus. *On Unbelievable Tales.* Translation, introduction, and commentary by Jacob Stern. Wauconda, IL: Bolchazy-Carducci Publishers, 1996. https://books.google.com/books?id=t4EfiGQwgh4C&printsec=frontcover&source=gbs_ge_summary_r&cad=0#v=onepage&q&f=false.

Penzias, A. A., and R. W. Wilson. "A Measurement of Excess Antenna Temperature at 4080 Mc/s." *Astrophysical Journal* 142 (1965): 419–421.

Perlmutter, S., et al. "Measurements of Ω and Λ from 42 High-Redshift Supernovas." *Astrophysical Journal* 517 (1999): 565–586.

Philoponus. *Against Aristotle, on the Eternity of the World.* Translated by Christian Wildberg. London: Gerald Duckworth, 1987.

———. *Philoponus: On Aristotle Physics 4.6–9.* Translated by Pamela Huby. London: Bloomsbury Academic, 2012.

Pirrie, George. *A Short Treatise of the General Laws of Motion and Centripetal Forces.* Edinburgh: William Adams Junior, 1720.

Plato. *The Republic of Plato.* Translated by Benjamin Jowett. Oxford: Clarendon Press, 1888. https://oll.libertyfund.org/titles/plato-the-republic-1888-ed.

———. *Symposium.* Translated by Benjamin Jowett. London: Pearson, n.d. http://classics.mit.edu/Plato/symposium.html.

Riess, Adam G., et al. "Observational Evidence from Supernovae for an Accelerating Universe and a Cosmological Constant." *Astronomical Journal* 16 (1998): 1009–1038.

————. "Type Ia Supernova Discoveries at $z > 1$ from the Hubble Space Telescope: Evidence for Past Deceleration and Constraints on Dark Energy Evolution." *Astrophysical Journal* 607 (2001): 665–687.

Rose, H. J. *A Handbook of Greek Mythology*. London: Methuen, 1958.

Rosen, Edward. "The Composition and Publication of Kepler's *Dream*." In *Kepler's Somnium: The Dream, or Posthumous Work on Lunar Astronomy*. Translated, with a commentary, by Edward Rosen. Mineola, NY: Dover Publications, 2003.

Sandage, Allan. "Cosmology: A Search for Two Numbers." *Physics Today* 23 (1970): 34–41.

Simplicius. *Simplicius: On Aristotle Physics 8.1–5*. Edited by Richard Sorabji. London: Bloomsbury Academic, 2012.

Smith, George. "Newton's *Philosophiae Naturalis Principia Mathematica*." In *The Stanford Encyclopedia of Philosophy*, Winter 2008. https://plato.stanford.edu/archives/win2008/entries/newton-principia/.

Sorabji, Richard. "The contra Aristotelem." In *Against Aristotle, on the Eternity of the World*, translated by Christian Wildberg. London: Gerald Duckworth, 1987a.

————. "John Philoponus." In *Philoponus and the Rejection of Aristotelian Science*, edited by Richard Sorabji. Ithaca, NY: Cornell University Press, 1987b.

————. "Preface." In *Philoponus and the Rejection of Aristotelian Science*, edited by Richard Sorabji. Ithaca, NY: Cornell University Press, 1987c.

Sorabji, Richard, ed. *Philoponus and the Rejection of Aristotelian Science*. Ithaca, NY: Cornell University Press, 1987.

————. *Simplicius: On Aristotle Physics 8.1–5*. London: Bloomsbury Academic, 2012.

Sprat, Thomas. *The History of the Royal Society of London, for the Improving of Natural Knowledge*. London: T.R., 1667. https://books.google.com/books?id=g3oOAAAAQAAJ&printsec=frontcover&source=gbs_ge_summary_r&cad=0#v=onepage&q&f=false.

Sproul, Barbara C. *Primal Myths: Creation Myths Around the World*. New York: Harper One, 1991.

Stewart, J. A. *The Myths of Plato*. London: Macmillan and Co., 1905.

Taylor, John H., ed. *Journey Through the Afterlife: Ancient Egyptian Book of the Dead*. Cambridge, MA: Harvard University Press, 2010.

Thorne, Kip S. *Black Holes and Time Warps: Einstein's Outrageous Legacy*. New York: W. W. Norton, 1994.

Voltaire (François-Marie Arouet). *Letters on the English*. New York: P. F. Collier & Son, 1909–14; Bartleby.com, 2001. www.bartleby.com/34/2/.

Waff, Craig B. "Comet Halley's First Expected Return." *Journal for the History of Astronomy* 17 (1986): 1–37.

Wallis, Ruth. "The Glory of Gravity — Halley's Comet 1759." *Annals of Science* 41 (1984): 279–286.

Walsh, D., R. F. Carswell, and R. J. Weymann. "0957 + 561 A, B: Twin Quasistellar Objects or Gravitational Lens?" *Nature* 279 (1979): 381–384.

West, Michael. "Einstein's Last Words." Lowell Observatory. https://lowell.edu/einsteins-last-words/.

Westfall, Richard S. *Never at Rest: A Biography of Isaac Newton*. Cambridge: Cambridge University Press, 1980.

INDEX

Index